T. Tachikawa, K. Nose, T. Ohmori, M. Adachi (Eds.)

New Trends in the Molecular and Biological Basis for Clinical Oncology

T. Tachikawa, K. Nose,
T. Ohmori, M. Adachi (Eds.)

New Trends in the Molecular and Biological Basis for Clinical Oncology

Tetsuhiko Tachikawa, D.D.S., Ph.D.
Professor and Chairman
Department of Oral Pathology and Diagnosis
Showa University School of Dentistry
1-5-8 Hatanodai, Shinagawa-ku, Tokyo 142-8555, Japan

Kiyoshi Nose, Ph.D.
Professor and Chairman
Department of Microbiology
Showa University School of Pharmaceutical Sciences
1-5-8 Hatanodai, Shinagawa-ku, Tokyo 142-8555, Japan

Tohru Ohmori, M.D., Ph.D.
Associate Professor
Institute of Molecular Oncology
Showa University
1-5-8 Hatanodai, Shinagawa-ku, Tokyo 142-8555, Japan

Mitsuru Adachi, M.D., Ph.D.
Professor and Chairman
Division of Allergology and Respiratory Medicine
Department of Internal Medicine
(First Department of Internal Medicine)
Showa University School of Medicine
1-5-8 Hatanodai, Shinagawa-ku, Tokyo 142-8666, Japan

Library of Congress Control Number: 2008937890

ISBN 978-4-431-88662-4 Springer Tokyo Berlin Heidelberg New York
e-ISBN 978-4-431-88663-1

Springer is a part of Springer Science+Business Media
springer.com

Printed in Japan

Typesetting: Camera-ready by the editors and authors
Printing and binding: Kato Bunmeisha, Japan

Printed on acid-free paper

Foreword

On behalf of Showa University, I am pleased to publish the proceedings of The 4th Annual Meeting of the Showa International Symposium for Life Sciences, *New Trends in the Molecular and Biological Basis for Clinical Oncology*.

This symposium aims for the improvement in the prevention and therapeutics for intractable diseases and has been held annually for the past four years. It always provides us with a great opportunity to share the latest knowledge in fundamental, translational, and clinical fields about intractable diseases. In last year's symposium we focused on medical oncology. The proceedings is composed of 18 articles presenting the latest data about new strategies for cancer therapeutics. Six specialists of the contributors are invited from all over the world.

I hope that this book will furnish valuable information for the development of clinical as well as basic cancer research to all who are working in this field.

Katsuji Oguchi, M.D., Ph.D.
Chairman, Board of Directors of Showa University
Tokyo, Japan
June 2008

Foreword

Recently, many cancer patients have experienced the hope of being treated with the latest cancer therapy by specialists in oncology. It is obvious, however, that the number of oncologists is absolutely inadequate to meet the demand nationwide, because medical oncology has not been authorized in the ministry's curriculum guideline and cancer therapeutics has not been included among the special subjects of medical schools in Japan. Accordingly, the establishment of an education system for oncology to produce many experts in cancer therapeutics is the prime task of medical schools. Consequently, many medical schools in Japan have begun to set up departments of clinical oncology in their hospitals, and we also have decided to establish such a department in our hospital. I believe that the work of clinical oncology should be further accelerated for the benefit of cancer patients in this country.

Especially under these circumstance, it is my great pleasure to publish *New Trends in the Molecular and Biological Basis for Clinical Oncology*, the proceedings of The 4th Annual Meeting of the Showa International Symposium for Life Sciences, which was held at Showa University on October 6th, 2007. On behalf of the university, I am grateful to all the contributors to this book, who are active leaders in clinical oncology and in cancer research.

Akiyoshi Hosoyamada, M.D., Ph.D.
President, Showa University
Tokyo, Japan
June 2008

Foreword

A long time has passed since cancer became the primary cause of death in Japan. In fact, the number of cancer patients is still growing, and one-third of clinical deaths are caused by this disease. It is estimated that in tandem with the rapid transition to an aging society, the increment in the number of cancer patients will accelerate. It is obvious that the conquest of cancer is still one of the largest medical problems to be solved.

Recently, remarkable progress in cancer therapy has been made possible by better understanding of the molecular and cell biology of cancer and by the technology that has resulted in the discovery of new drugs. The molecular target cancer drugs are being developed constantly and with astonishing speed, and at the same time many large-scale clinical studies are being conducted all over the world. In order to utilize these cancer drugs properly, the efficacy of each of them must be proved by their own clinical studies. However, evaluation in clinical studies is not sufficient to select the optimal drug for each individual patient, because the impact of the molecular target drugs is dependent on the "individuality" of cancer. For this purpose, we have to solve the following additional problems:

1) Proof that a molecular target drug can actually inhibit the expected target in vivo (proof of principle)
2) Detection of a clinical marker from patient materials correlated with drug efficacy (surrogate marker)

As these challenges can be met only through close cooperation between basic and clinical cancer research, I think that we should establish an efficient new research system, breaking down the barrier between those two areas of research. There is only cancer research for the benefit of patients. We must be united as one body of researchers, wishing to cure cancer.

Hajime Yasuhara, M.D., Ph.D.
Dean, Showa University School of Medicine
Tokyo, Japan
June 2008

Preface

Several years ago, molecular target anticancer drugs began to be used in the clinical field, showing outstanding clinical efficacy. These newly developed drugs have a great effect even on intractable cancers such as hepatoma and renal cell carcinoma. Fortunately, we are now facing a surge of new findings in the development of cancer therapy. Cancer research is becoming a multidisciplinary field that includes nearly all areas of science and technology. In fact, many anticancer drugs are developed by basic cancer research that is increasingly influenced by clinical observations. At the same time, the molecular and genetic observations of epidemiology have great influence on research protocols in both the laboratory and the clinic. We are convinced that such cooperation across basic, translational, and clinical research fields will rapidly advance cancer therapeutics beneficial to cancer patients.

Under these circumstances, we held The 4th Annual Meeting of Showa International Symposium for Life Sciences with the theme "New Trends in the Molecular and Biological Basis for Clinical Oncology" at Showa University, Tokyo, in October 2007. This book contains the materials from that symposium in which new strategies for the development of cancer therapeutics were discussed. In the first chapter, we focus on the cancer therapy targeting receptor tyrosine kinases aimed at clinical utilization of new signaling regulations. The second chapter explains the interaction between cancer progression and extracellular environments such as inflammatory cytokines and the extracellular matrix. The third describes investigation of the biomarkers for personalized cancer therapy, using microarray analysis and pharmacogenomics technology. This book also includes several reports of the latest investigations dealing with cancer cell biology and therapeutics.

We thank all the authors and others who have contributed to the publication of this book. We believe that its contents are valuable for investigators in cancer research and that their achievements will help to provide the best possible care for individual cancer patients in the near future.

Tetsuhiko Tachikawa, Professor of Oral Pathology
Kiyoshi Nose, Professor of Microbiology
Tohru Ohmori, Associate Professor of Molecular Oncology
Mitsuru Adachi, Professor of Internal Medicine
Showa University, Tokyo, Japan
June 2008

Acknowledgments

The 4th Annual Meeting of the Showa International Symposium for Life Sciences was supported in part by Grants for the Promotion of the Advancement of Education and Research in Graduate Schools and Ordinary Expenses for Private Schools from the Ministry of Education, Culture, Sports, Science and Technology of Japan.

We would like to acknowledge the extraordinary contributions of Mrs. Noriko Shishido, who played a vital role in the preparation of this book and assumed responsibility for the organization and compilation of all the chapters. We are grateful for her dedication.

TT
KN
TO
MA

Contents

Part III. Biomarkers for Clinical Oncology

Part IV. Poster Session

Contributors

Adachi, M., First Department of Internal Medicine, School of Medicine, Showa University, Hatanodai 1-5-8, Shinagawa-ku, Tokyo 142-8666, Japan

Ando, K., First Department of Internal Medicine, School of Medicine, Showa University, Hatanodai 1-5-8, Shinagawa-ku, Tokyo 142-8666, Japan

Arao, T., Department of Genome Biology, Kinki University School of Medicine, 377-2 Ohnohigashi, Osakasayama, Osaka 589-8511, Japan

Arteaga, C.L., Department of Cancer Biology and Medicine and Vanderbilt-Ingram Cancer Center, Vanderbilt University School of Medicine, Nashville, TN 37232, USA

Bornstein, S., Departments of Otolaryngology, Cell & Developmental Biology, and Dermatology, Oregon Health & Science University, Portland, OR, USA

Chen, H.-Y., Institute of Statistical Science, Academia Sinica, 128, Sec. 2, Academia Road, Taipei 115, Taiwan

Egawa, K., Department of Microbiology, Showa University School of Pharmaceutical Sciences, Hatanodai 1-5-8, Shinagawa-ku, Tokyo 142-8555, Japan

Fujiwara, T., Division of Surgical Oncology, Department of Surgery, Okayama University, Graduate School of Medicine and Dentistry, Okayama, Japan; Center for Gene and Cell Therapy, Okayama University Hospital, 2-5-1 Shikata-cho, Okayama 700-8558, Japan

Han, G.W., Departments of Otolaryngology, Cell & Developmental Biology, and Dermatology, Oregon Health & Science University, Portland, OR; Departments of Pathology and Otolaryngology, University of Colorado Denver, Aurora, CO, USA

Hara, S., Department of Health Chemistry, School of Pharmaceutical Sciences, Showa University, 1-5-8 Hatanodai, Shinagawa-ku, Tokyo 142-8555, Japan

Hasebe, Y., Department of Microbiology, Showa University School of Pharmaceutical Sciences, Hatanodai 1-5-8, Shinagawa-ku, Tokyo 142-8555, Japan

Hatori, M., Department of Oral and Maxillofacial Surgery, School of Dentistry, Showa University, 2-1-1 Kitasenzoku, Ohta-ku, Tokyo 145-8515, Japan

Hirose, T., First Department of Internal Medicine, School of Medicine, Showa University, Hatanodai 1-5-8, Shinagawa-ku, Tokyo 142-8666, Japan

Hoot, K., Departments of Otolaryngology, Cell & Developmental Biology, and Dermatology, Oregon Health & Science University, Portland, OR, USA

Horichi, N., First Department of Internal Medicine, Showa University School of Medicine, 1-5-8 Hatanodai, Shinagawa-ku, Tokyo, 142-8666, Japan

Hosaka, T., First Department of Internal Medicine, School of Medicine, Showa University, Hatanodai 1-5-8, Shinagawa-ku, Tokyo 142-8555, Japan

Inoue, F., Institute of Molecular Oncology, Showa University, Hatanodai 1-5-8, Shinagawa-ku, Tokyo 142-8555, Japan

Irié, T., Department of Oral Pathology and Diagnosis, Showa University School of Dentistry, Hatanodai 1-5-8, Shinagawa-ku, Tokyo 142-8555, Japan

Ishikawa, F., Department of Microbiology, Showa University School of Pharmaceutical Sciences, Hatanodai 1-5-8, Shinagawa-ku, Tokyo 142-8555, Japan

Isobe, T., Department of Oral Pathology and Diagnosis, Showa University School of Dentistry, Hatanodai 1-5-8, Shinagawa-ku, Tokyo 142-8555, Japan

Ito, D., Department of Oral and Maxillofacial Surgery, School of Dentistry, Showa University, 2-1-1 Kitasenzoku, Ohta-ku, Tokyo 145-8515, Japan

Kadofuku, T., Institute of Molecular Oncology, Showa University, Hatanodai 1-5-8, Shinagawa-ku, Tokyo 142-8555, Japan

Kanome, T., Institute of Molecular Oncology, Showa University, 1-5-8 Hatanodai, Shinagawa-ku, Tokyo 142-8555, Japan

Kojima, T., Division of Surgical Oncology, Department of Surgery, Okayama University, Graduate School of Medicine and Dentistry, Okayama, Japan; Center for Gene and Cell Therapy, Okayama University Hospital, 2-5-1 Shikata-cho, Okayama 700-8558, Japan

Kurihara, Y., Department of Oral and Maxillofacial Surgery, School of Dentistry, Showa University, 2-1-1 Kitasenzoku, Ohta-ku, Tokyo 145-8515, Japan

Kuroki, T., Gifu University, 1-1 Yanagido Gifu-shi, Gifu 501-1194, Japan

Kusumoto, S., First Department of Internal Medicine, Showa University School of Medicine, 1-5-8 Hatanodai, Shinagawa-ku, Tokyo 142-8666, Japan

Lu, S.L., Departments of Otolaryngology, Cell & Developmental Biology, and Dermatology, Oregon Health & Science University, Portland, OR; Departments of Pathology and Otolaryngology, University of Colorado Denver, Aurora, CO, USA

Matsunaga, T., Department of Oral Pathology and Diagnosis, Showa University School of Dentistry, Hatanodai 1-5-8, Shinagawa-ku, Tokyo 142-8555, Japan

Miller, T., Department of Medicine, Vanderbilt University School of Medicine, Nashville, TN 37232, USA

Mori, K., Department of Microbiology, Showa University School of Pharmaceutical Sciences, 1-5-8 Hatanodai, Shinagawa-ku, Tokyo 142-8555, Japan

Nagumo, T., Department of Oral and Maxillofacial Surgery, School of Dentistry, Showa University, 2-1-1 Kitasenzoku, Ohta-ku, Tokyo 145-8515, Japan

Nakashima, M., First Department of Internal Medicine, Showa University School of Medicine, 1-5-8 Hatanodai, Shinagawa-ku, Tokyo 142-8666, Japan

Nishio, K., Department of Genome Biology, Kinki University School of Medicine, 377-2 Ohnohigashi, Osakasayama, Osaka 589-8511, Japan

Nose, K., Department of Microbiology, Showa University School of Pharmaceutical Sciences, 1-5-8 Hatanodai, Shinagawa-ku, Tokyo 142-8555, Japan

Ohmori, T., Institute of Molecular Oncology, Showa University, Hatanodai 1-5-8, Shinagawa-ku, Tokyo 142-8555, Japan

Ohnishi, T., First Department of Internal Medicine, Showa University School of Medicine, 1-5-8 Hatanodai, Shinagawa-ku, Tokyo 142-8666, Japan

Okuda, K., First Department of Internal Medicine, Showa University School of Medicine, 1-5-8 Hatanodai, Shinagawa-ku, Tokyo 142-8666, Japan

Oshima, Y., Department of Microbiology, Showa University School of Pharmaceutical Sciences, 1-5-8 Hatanodai, Shinagawa-ku, Tokyo 142-8555, Japan

Polk, D.B., Department of Pediatrics, Cell and Developmental Biology, Vanderbilt University School of Medicine, 1025 MRBIV, Nashville, TN 37232-0696, USA

Saijo, N., Internal Medicine, National Cancer Center Hospital East, 6-5-1 Kashino-ha Kashiwa-shi Chiba 277-8577, Japan

Sasaki, Y., Department of Medical Oncology, Saitama International Medical Center – Comprehensive Cancer Center, Saitama Medical University, 1397-1 Yamane, Hidaka, Saitama 350-1298, Japan

Seiki, M., Division of Cancer Cell Research, Institute of Medical Science, University of Tokyo, 4-6-1 Shirokane-dai, Minato-ku, Tokyo 108-8639, Japan

Shibanuma, M., Department of Microbiology, Showa University School of Pharmaceutical Sciences, 1-5-8 Hatanodai, Shinagawa-ku, Tokyo 142-8555, Japan

Shin, I., Department of Cancer Biology, Vanderbilt University School of Medicine, Nashville, TN 37232, USA; Department of Life Science, Hanyang University, Seoul 133-791, Korea

Shintani, S., Department of Oral and Maxillofacial Surgery, School of Dentistry, Showa University, 2-1-1 Kitasenzoku, Ohta-ku, Tokyo 145-8515, Japan

Shirai, T., First Department of Internal Medicine, Showa University School of Medicine, 1-5-8 Hatanodai, Shinagawa-ku, Tokyo 142-8666, Japan

Shirota, T., Department of Oral and Maxillofacial Surgery, School of Dentistry, Showa University, 2-1-1 Kitasenzoku, Ohta-ku, Tokyo 145-8515, Japan

Sugimoto, T., Department of Microbiology, Showa University School of Pharmaceutical Sciences, Hatanodai 1-5-8, Shinagawa-ku, Tokyo 142-8555, Japan

Sugiyama, T., First Department of Internal Medicine, Showa University School of Medicine, 1-5-8 Hatanodai, Shinagawa-ku, Tokyo 142-8666, Japan

Tachikawa, T., Department of Oral Pathology and Diagnosis, Showa University School of Dentistry, Hatanodai 1-5-8, Shinagawa-ku, Tokyo 142-8555, Japan

Urata, Y., Oncolys BioPharma, Inc., 3-16-33 Roppongi, Minato-ku, Tokyo 106-0032, Japan

Wakamatsu, M., Department of Microbiology, Showa University School of Pharmaceutical Sciences, Hatanodai 1-5-8, Shinagawa-ku, Tokyo 142-8555, Japan

Wang, E.S., Department of Cancer Biology, Vanderbilt University School of Medicine, Nashville, TN 37232, USA

Wang, X.-J., Departments of Otolaryngology, Cell & Developmental Biology, and Dermatology, Oregon Health & Science University, Portland, OR; Departments of Pathology and Otolaryngology, University of Colorado Denver, Aurora, CO, USA

Watanabe, Y., Center for Gene and Cell Therapy, Okayama University Hospital, 2-5-1 Shikata-cho, Okayama 700-8558, Japan; Oncolys BioPharma, Inc., 3-16-33 Roppongi, Minato-ku, Tokyo 106-0032, Japan

Yamamoto, G., Department of Oral Pathology and Diagnosis, Showa University School of Dentistry, Hatanodai 1-5-8, Shinagawa-ku, Tokyo 142-8555, Japan

Yamaoka, T., First Department of Internal Medicine, Showa University School of Medicine, 1-5-8 Hatanodai, Shinagawa-ku, Tokyo 142-8666, Japan

Yang, P.-C., Department of Internal Medicine, National Taiwan University Collage of Medicine, 1, Sec 1, Ren-Ai Rd, Taipei 100, Taiwan

Yoshiba S., Department of Oral and Maxillofacial Surgery, School of Dentistry, Showa University, 2-1-1 Kitasenzoku, Ohta-ku, Tokyo 145-8515, Japan

Color Plates

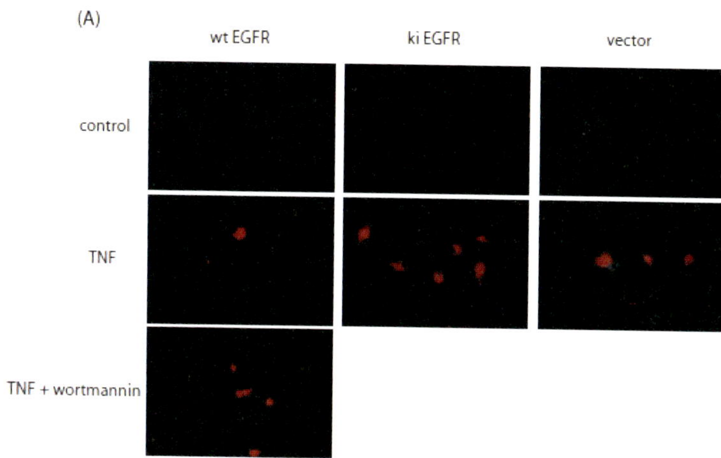

Part I: Yamaoka et al (Fig.1A)

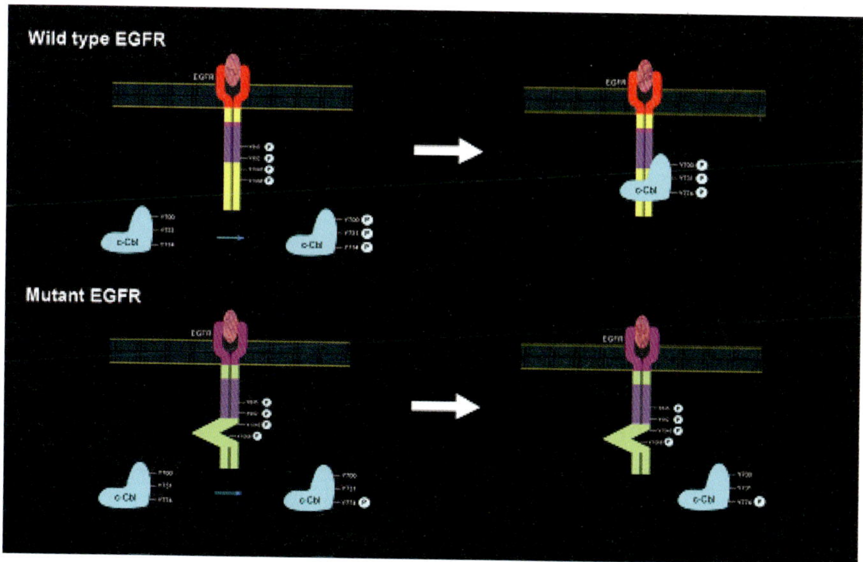

Part I: Ohmori et al (Fig.6)

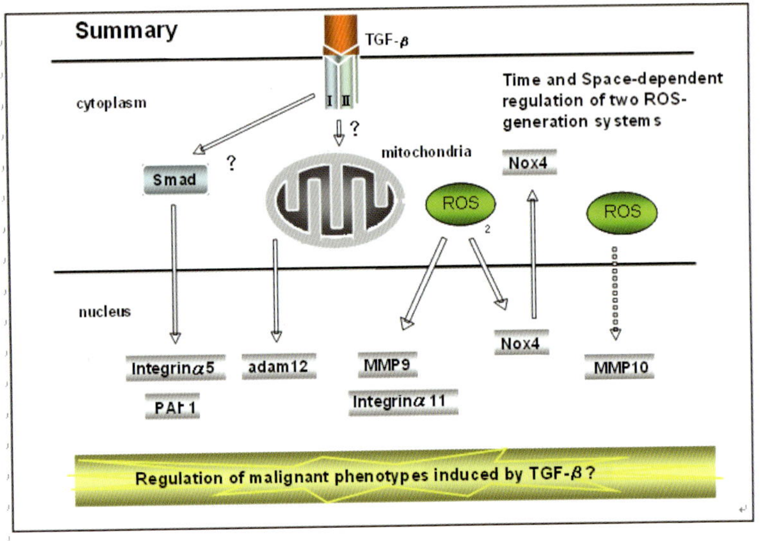

Part II: Nose et al (Fig.3)

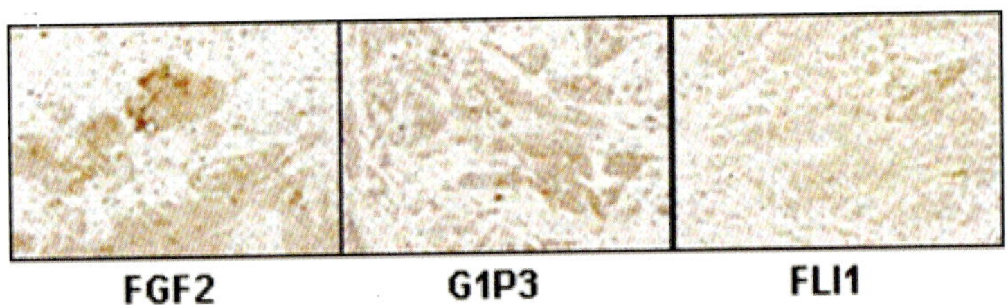

Part III: Shintani et al (Fig.3)

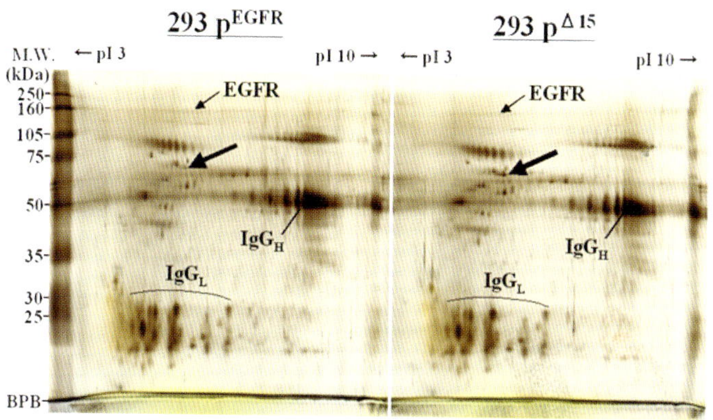

Part IV: Kadofuku et al (Fig.1)

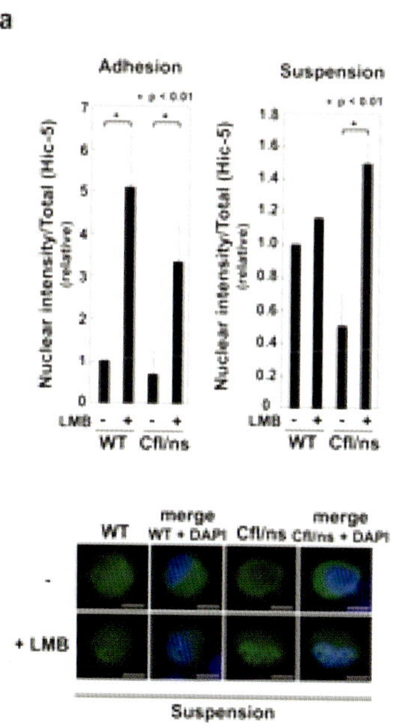

Part IV: Mori et al (Fig.3a)

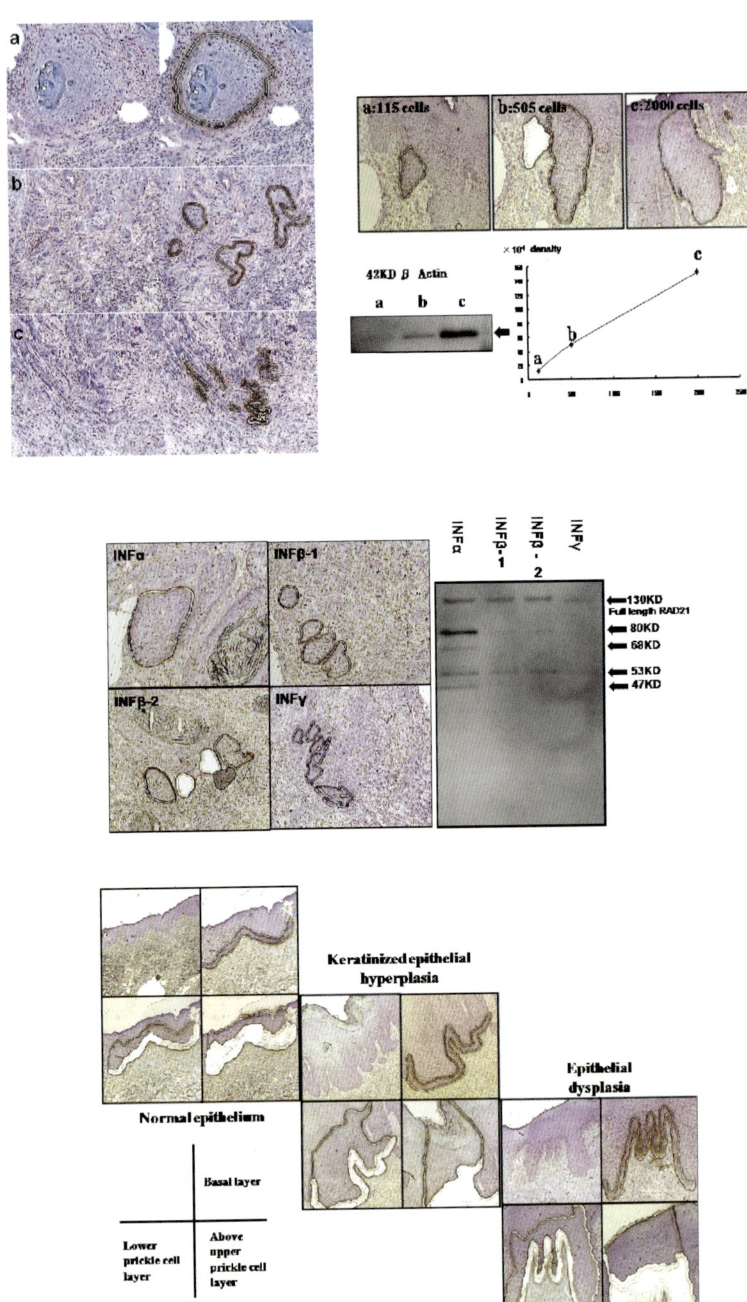

Part IV: Yamamoto et al (Figs.1, 3-5)

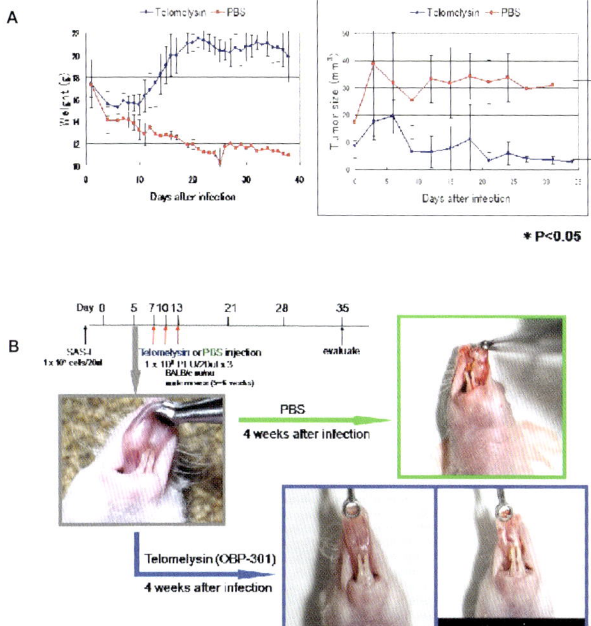

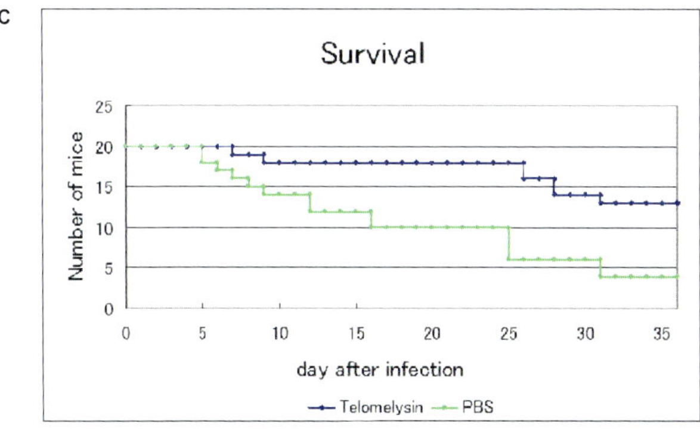

Part IV: Kurihara et al (Fig.6)

Part I

RTK-Related Signaling

HER2 Overexpression Attenuates the Antiproliferative Effect of Aromatase Inhibitor in MCF-7 Breast Cancer Cells Stably Expressing Aromatase

Incheol Shin[1,2], Todd Miller[3], E. Shizen Wang[1] and Carlos L. Arteaga.[1,3,4*]

[1] Department of Cancer Biology and [3]Medicine and [4]Vanderbilt-Ingram Cancer Center, Vanderbilt University School of Medicine, Nashville,Tennessee 37232
[2]Department of Life Science, Hanyang University, Seoul 133-791 Korea

Summary. We investigated the effect of HER2 overexpression on resistance to aromatase inhibitor letrozole in MCF-7 breast cancer cells stably expressing cellular aromatase (MCF-7/CA). Forced expression of HER2 in MCF-7/CA cells by infection with retroviral vector showed increased Akt and MAPK phosphorylation and phosphorylation of estrogen receptor (ER) α on Ser118 and Ser167. MCF-7/CA cells overexpressing HER2 (MCF-7/CA pBMN-HER2) showed more than 2 fold higher level of ER-mediated transcription activity upon androstenedione treatment as compared to vector control MCF-7/CA (MCF-7/CA pBMN-vec) cells and letrozole co-treatment did not effectively block androstenedione-induced transcription. Inhibition of Akt and MAPK by LY294002 and UO126 in MCF-7/CA pBMN-HER2 cells, respectively, resulted in the reduction of androstenedione-induced transcription activities comparable to those of MCF-7/CA pBMN-vec cells. MCF-7/CA pBMN-HER2 cells showed increased proliferation in the medium containing charcoal-stripped serum supplemented with androstenedione as compared to MCF-7/CA pBMN-vec cells. Moreover, co-treatment of letrozole could not efficiently abrogate androstenedione-mediated cell proliferation in MCF-7/CA pBMN-HER2 cells. MCF-7/CA pBMN-HER2 cells were also shown to be resistant to letrozole-induced apoptosis. Aromatase activities of both MCF-7/CA pBMN-HER2 and MCF-7/CA pBMN-vec cells were reduced to the similar basal levels by letrozole treatment, suggesting that resistance to letrozole is not conferred by the resistance of aromatase enzyme itself to letrozole. Chromatin immunopreciptation assays revealed that MCF-7/CA pBMN-HER2 cells showed ligand-independent association of estrogen receptor (ER) α, coactivators AIB1 and CBP to the promoter region of pS2 gene. Upon androstenedione treatment, there were increased associations of ERα, AIB1 and CBP with pS2 promoter in MCF-7/CA pBMN-HER2 cells as compared to MCF-7/CA pBMN-vec cells. These results suggest that recruitment of coactivator complexes to estrogen-responsive promoter by HER2 overexpression may play a role in developing letrozole resistance through upregulation of estrogen-responsive genes and increased cell survival and proliferation.

Key words. Aromatase, HER2, Breast Cancer, letrozole, estrogen receptor

Introduction

Aromatase, a cytochrome P-450 hemoprotein, is the product of *CYP19* gene and expressed in a wide variety of tissues including the placenta, ovary, brain, bone and breast (Bulun et al. 1997, Means et al. 1991, Naftolin et al. 1975, Shozu and Simpson 1998, Perel et al. 1982, Silva et al. 1989). Aromatase is a cytochrome P450 enzyme that catalyzes conversion of testosterone and androstenedione (AD) to estradiol (E2) and estrone, respectively (Long et al. 1998, Seralini and Moslemi 2001) playing a key role in estrogen biosynthesis. The expression of aromatase is elevated in breast tumor tissue as compared with normal breast tissue, suggesting an important role of aromatase in breast tumorigenesis (Bulun et al. 1993).

Approximately 70 % of breast cancers are estrogen receptor (ER) positive, so antiestrogen therapy has been an important treatment for breast cancer. Tamoxifen has been the drug of choice for hormone-dependent breast cancer patients (Litherland and Jackson 1988), but tamoxifen's partial estrogenic activity is believed to mediate resistance to the drug and relapse of patients (Leonessa et al. 1992). Unlike tamoxifen, the non-steroidal aromatase inhibitors do not show estrogenic properties and show superior efficacy as compared to tamoxifen (Mouridsen et al. 2001). Although ovary is no longer producing estrogens in menopausal women, estrogens are still synthesized from androgens via aromatizaton catalyzed by aromatase in peripheral tissues (Hemsell et al. 1974). So it is believed that inhibition of *de novo* estrogen synthesis by aromatase inhibitors would be an alternative treatment to post-menopausal breast cancer patients. The aromatase inhibitors, letrozole, anastrozole and exemestane were recently used for the second-line therapies for tamoxifen-relapased breast cancer (Dombernowsky et al. 1998, Buzdar et al. 1998, Kaufmann et al. 2000, Long et al. 2002).

The HER2 receptor is the product of the *HER2* proto-oncogene and a member of the EGFR (HER1) family of receptor tyrosine kinases, including HER3 and HER4. These family of receptors consist of cysteine-rich extracellular domain, a single membrane-spanning domain, and a cytoplasmic domain including a tyrosine kinase domain (Riese and Stern 1998, Olayioye et al. 2000). The ligands bind the extracellular domain of these receptors leading to the formation of both homo- and heterodimers. Subsequently, dimerization triggers autophosphorylation of the tyrosine residues of the receptors in their cytoplasmic domains and activates signaling pathways that enhances cell proliferation and survival (Olayioye et al. 2000). Though no direct ligand for HER2 has been discovered, HER2 was shown to serve as a preferred heterodimerization partner for all other HER family members of receptor tyrosine kinase (Tzahar et al. 1997, Graus-Porta et al. 1997). Elevated levels of HER2 were found in 20 – 30 % of human breast cancer because of gene amplification of *HER2* proto-oncogene (Slamon et al. 1989). This gene amplification of *HER2* significantly correlates with negative clinical prognosis (Andrulis et al. 1998). Both experimental and clinical evidence suggest a correlation between overexpression and/or aberrant activity of the HER2/neu (erbB2) signaling pathway and antiestrogen resistance in breast cancer (Benz et al. 1992, Liu et al. 1995, Yamauchi et al. 1997, Houston et al. 1999). HER2 overexpression results in activation of Ras/MAPK signaling in breast tumor cells (Tzahar and Yarden 1998, Janes et al. 1994), which may contribute to tumor proliferation regardless of estrogen receptor (ER) signaling. Since efficacy of tamoxifen is compromised in HER2 positive breast cancer, it is intriguing to check the effect of HER2 on estrogen deprivation by aromatase inhibitor in breast cancer cells. Therefore, we investigated the effect of HER2 overexpression on resistance to aromatase inhibitor letrozole using MCF-7 breast cancer cells stably transfected with cellular aromatase (MCF-7/CA). Forced expression of HER2 led to decreased inhibitory effect of letrozole on ER-mediated transcription and cancer cell proliferation, providing a rationale for a possible new approach for combined targeting therapy.

Materials and Methods

Generation of MCF-7/CA pBMN-HER2 cells

MCF-7/CA cells with tetracycline-controllable aromatase expression were obtained from Dr. Yue Wei (Virginia University) and maintained in IMEM with 10 % FBS, 200 μg/ml G418 and 100 μg/ml hygromycin. Generation of retroviral vector pBMN-HER2-IRES-EGFP and production of retrovirus were performed as described (Ueda et al. 2004). MCF-7/CA cells were transduced with vector control retrovirus or HER2 retrovirus for 3 hours in the presence of polybrene (6 μg/ml). After 3 passages, cells stably expressing EGFP were sorted by flow cytometry.

Immunoprecipitation and immunoblot analyses

The cell lysates were prepared as described elsewhere (Kurokawa et al. 2000). Fifty μg of protein were resolved by SDS-PAGE and subjected to immunoblot analysis with antibodies against HER2 (Neomarkers, Freemont, CA), Akt, p-Ser 473 Akt, MAPK, ERα, p-Ser118 ERα, p-Ser167 ERα (Cell Signaling, Beverly, MA), p-MAPK (Promega, Madison, WI). Immunoprecipitations for phosphor-HER2 analysis were carried out by incubating 0.5 mg of total cell lysates with 1 μg of HER2 antibody at 4°C overnight. Protein A-Sepharose (Sigma, St Louis, MO; 1:1 slush in PBS) was then added for 2 h at 4°C while rocking. The precipitates were washed four times with ice-cold PBS, resuspended in 6x Laemmli sample buffer, and resolved by SDS-PAGE followed by immunoblot analysis with phosphotyrosine antibody (Sigma).

Luciferase ER reporter assay

The pGLB-MERE plasmids were provided by Dr. El-Ashry (Georgetown University, Washington D.C.) The plasmid contains double consensus ERE. MCF-7/CA pBMN-vec or MCF-7/CA pBMN-HER2 cells in 12-well plates were transiently transfected with 0.5 μg/well of pGLB-MERE and 0.0025 μg/ml of pCMV-Rl (Promega) using FuGENE 6 (Roche, Indianapolis, IN) transfection reagent. Eighteen h later, the medium was changed with in a phenol-red free IMEM supplemented with 10 % charcoal stripped serum (Hyclone, Logan, UT) and incubated for further 24 h. The cells were then treated with various combinations of 25 nM AD (Sigma), 100 nM letrozole

(Novartis, Basel, Switzerland), 1 μM E2 (Sigma), 20 μM LY294002 (EMB Biosciences, Darmstadt, Germany), 5 μM U0126 (EMB Biosciences) and 3 μM ZD1839 (Novartis) as indicated for additional 24 h. Firefly luciferase (Luc) and *Renilla reniformis* luciferase activities (RlLuc) in cell lysates were determined using the Dual Luciferase Reporter Assay System (Promega) according to the manufacturer's protocol in a Monolight 2010 luminometer (Analytical Luminescence Laboratory, San Diego, CA). Luc activity was normalized to RlLuc activity and presented as Relative Luciferase Units. All assays were done in triplicate wells.

Proliferation assay

Both MCF-7/CA pBMN-vec and MCF-7/CA pBMN-HER2 cells were seeded at the density of 5×10^5 cells in T25 flasks. After 2 days, the medium was replaced with phenol-red free IMEM containing 10 % charcoal stripped-serum and maintained for 3 more days until seeded at the density of 1×10^4 cells per well in 12 well-plates. After 2 more days, the cells were treated with hormones and drugs as indicated and treated with fresh medium containing hormones and drugs for 2 times more on day 4 and day 6 after the cell seeding. On day 8 after the cell seeding onto 12 well-plates, the cells were harvested with trypsin and cell numbers were counted using a Coulter counter (Coulter electronics, Hialeah, IL).

Apoptosis assays

Cells ($5x10^5$/well) were seeded in 6-well plates in triplicate in serum-containing medium. The following day, the medium was changed to phenol red-free medium supplemented with 10 % charcoal-stripped serum containing 25 nM AD with or without 100 nM letrozole and incubated for further 4 days. After 2 days, the medium was replaced with a fresh medium containing AD or AD and letrozole. Both floating cells and adherent cells were collected 4 days after the initial treatment. Pooled cells were washed with PBS and then subjected to TUNEL analysis with the use of an Apo-BrdU assay kit (Phoenix Flow Systems, San Diego, CA) according to the manufacturer's protocol. TUNEL+ cells were quantitated in a FACS/Calibur Flow Cytometer (BD Biosciences, Mansfield, MA). For the caspase-3 cleavage assay, the cells were treated with 25 nM AD ± 100 nM letrozole for 4 days in a pheonol red-free IMEM containing charcoal-stripped serum. After 2 days, the medium was replaced with fresh medium containing AD or AD and letrozole. The cells lysates were prepared as described (Kurokawa et al. 2000), resolved on 12 % SDS-PAGE and followed by immunoblot analysis with caspase-3 antibody (Cell Signaling).

Aromatase assay

Aromatase assays were performed as described (Yue et al. 2003) with slight modifications. Briefly, the cells were seeded in 6-well plates at the density of 3×10^5 cells/well and incubated with 1 ml of IMEM supplemented with 10 % charcoal stripped-serum containing 1 μCi/ml AD (specific activity 25.3 Ci/mmol) (Perkin Elmer Lifesciences, Boston, MA) for 3 h. After incubation the medium was transferred to a test tube and 0.3 ml of 20 % tricholoroacetic acid was added to precipitate proteins. After centrifugation at $2{,}000 \times g$ for 15 min, 0.5 ml of supernatant was mixed with 1 ml of chloroform , vortex mixed and centrifuged at $2{,}000 \times g$ for 15 min. After centrifugation, 0.35 ml of aqueous phase was treated with 0.35 ml of 2.5 % activated charcoal suspension for 1 h. After brief centrifugation, aromatization activity was deterimend by measuring radioactivities in the supernatnant by scintillation counter (Beckman, Fullerton, CA). The aromatase activities were normalized with protein mass of the cells in 6-well plates.

ChIP assay

Cells were grown on T125 flask to a 60% confluence (about 10^7 cells) on a 150 mm dish. The cells were then incubated in a phenol red-free IMEM supplemented with 10 % charcoal-stripped medium for 3 days and serum starved for additional 24 h. The cells were pretreated for ZD1839 (3 μM) or vehicle (DMSO) for 3 h and treated with vehicle (ethanol) or AD (25 nM) with our without letrozole (100 nM) for 1 h. The cells were then washed with PBS and cross-linked with 1% formaldehyde in PBS at 37°C for 10 min. The cross-linking was stopped by adding 1 M Glycine to a final concentration of 125 mM and incubating at room temperature for 5 min. The cell monolayer was then washed twice with ice-cold PBS and scraped into 1 ml of ice-cold PBS containing protease inhibitors. After collecting the cells by centrifugation, the cell pellet was resuspended in 0.3 ml of Lysis Buffer (50 mM Tris-HCl [pH 8.1], 10 mM EDTA, 1% SDS, plus protease inhibitors (Roche) and 20 mM

NaF, 1 mM Na_3VO_4), and incubated on ice for 10 min. Sonication was then performed on ice at 7 watts, 20 sec each time for 3 times to shear the chromatins to fragments ranging from 300 to 1000 bp. After centrifugation, 20 μl of the supernatant was set aside as the input chromatin fraction, and the rest sample was 1:5 diluted in Dilution Buffer (20 mM Tris-HCl [pH 8.1], 150 mM NaCl, 2 mM EDTA, 0.01% SDS, 1% Triton X-100, plus protease inhibitors (Roche) and 20 mM NaF, 1 mM Na_3VO_4). To preclear the sample, 2 μg sonicated salmon sperm DNA, 2 μg BSA, 5 μg normal mouse IgG and 50 μl protein A/G-sepharose (a 50% slurry of 1:1 mixed protein A/G-sepharose in Dilution Buffer) were added and rotated for 2 h at 4°C. Precleared chromatin was aliquotated in half; 5 μg ERα antibody (Santa Cruz Biotechnology, Santa Cruz, CA), 5 μg of AIB1 antibody (Affinity bioreagents, Golden, CO) or 5 μg CBP antibody (Neomarkers) were added into either aliquot, and rotated overnight at 4°C, followed by addition of 50 μl protein A/G-sepharose, 2 μg of salmon sperm DNA, 2 μg BSA and another hour incubation at 4°C. Sepharose beads were harvested by centrifugation and washed sequentially in 1 ml of Low Salt Wash Buffer (20 mM Tris-HCl [pH 8.1], 150 mM NaCl, 2 mM EDTA, 0.1% SDS, 1% Triton X-100), High Salt Wash Buffer (20 mM Tris-HCl [pH 8.1], 500 mM NaCl, 2 mM EDTA, 0.1% SDS, 1% Triton X-100), LiCl Wash Buffer (0.25 M LiCl, 10 mM Tris-HCl [pH 8.1], 1 mM EDTA, 1% NP-40, 1% deoxycholate), and TE Buffer (10 mM Tris-HCl [pH 8.0], 1 mM EDTA) (twice) on a rotating platform with 10 min intervals. Protein-chromatin complexes were eluted from the beads in 100 μl Elution Buffer (0.1 M $NaHCO_3$, 1% SDS) by rotating at room temperature for 15 min. Then, 10 μg RNase and 6 μl of 5 M NaCl were added to the samples and incubated at 65°C overnight to reverse the formaldehyde cross-links. The cross-links in the input chromatin fraction were reversed as well. Twenty μl 1 M Tris-HCl [pH 6.5], 10 μl of 0.5 M EDTA and 20 μg Proteinase K were then added to each sample, followed by an incubation at 45°C for 2 h. DNA were extracted by Qiaquick PCR purification kit (Qiagen, Valencia, CA), and resuspended in 20 μl TE Buffer (pH 8.0). PCR were performed using 5 μl ChIP samples as templates and specific primers for the promoter region of PS2 gene: 5'-GGC CAT CTC TCA CTA TGA ATC ACT TCT GCA-3' (forward) and 5'-GGC AGG CTC TGT TTG CTT AAA GAG CGT TAG ATA-3' (reverse). After 30 cycles of amplification with the following settings for the thermocycler: 30 sec of denaturing at 94 ^{0}C, annealing for 30 sec at 63 ^{0}C and elongation for 1 min at 72 ^{0}C. The PCR products were analyzed on a 2% agarose gel.

Inoculation of MCF-7/CA cells to ovaryectomized athymic mice

Female ovaryectomized BALB/c athymic mice were purchased from Harlan and housed in a pathogen-free environment. Subconfluent cultures of both MCF-7/CA pBMN-vec and MCF-7/CA pBMN-HER2 cells were scraped into IMEM supplemented with 10 % FBS, centrifuged and resuspended in IMEM with 10 % FBS. Cells in suspension was inoculated subcutaneously (sc) in 0.1 ml of the cell suspension (2-5 $\times$ 10^7 cells/ml). The animals were then injected daily with 0.1 ml of 1 mg/ml in 0.3 % carboxymethylcellulose solution. After 60 days when the tumors reached measurable size, the mice were treated daily with 0.1 ml of 1 mg/ml in 0.3 % carboxymethylcellulose solution $\pm$ 0.1 ml of 0.6 mg/ml letrozole in the same solution. After 5 more days the mice were sacrificed and the tumors were harvested.

Results

Expression of HER2 in MCF-7/CA cells increases Akt and MAPK signaling

To generate a stable MCF-7/CA overexpressing HER2, a retroviral vector pBMN-IRES-EGFP was used to deliver HER2 into MCF-7/CA cells (MATERIALS AND METHODS). The expression of transgene was confirmed by flow cytometry as shown in Fig. 1A. Both vector control cells (MCF-7/CA pBMN-vec) and HER2 overexpressing cells (MCF-7/CA pBMN-HER2) showed around 100 fold higher green fluorescence as compared to parental cells (MCF-7/CA) suggesting that EGFP is expressed from the polycistronic transcripts containing both HER2 and EGFP. The expression of HER2 was also confirmed by western blots in Fig. 1B. As compared to parental MCF-/CA and MCF-7/CA pBMN-vec cells, MCF-7/CA pBMN-HER2 cells showed overexpression of HER2 and it is phosphorylated on tyrosine residues as determined by phosphotyrosine immunoblot. Forced expression of HER2 could increase the phosphorylation of Akt and MAPK. The phosphorylation of ER α on Ser118 and Ser167, a MAPK (Kato et al.

1995) and Akt (Campbell et al. 2001) phosphorylation sites, respectively, was also increased in MCF-7/CA pBMN-HER2 cells.

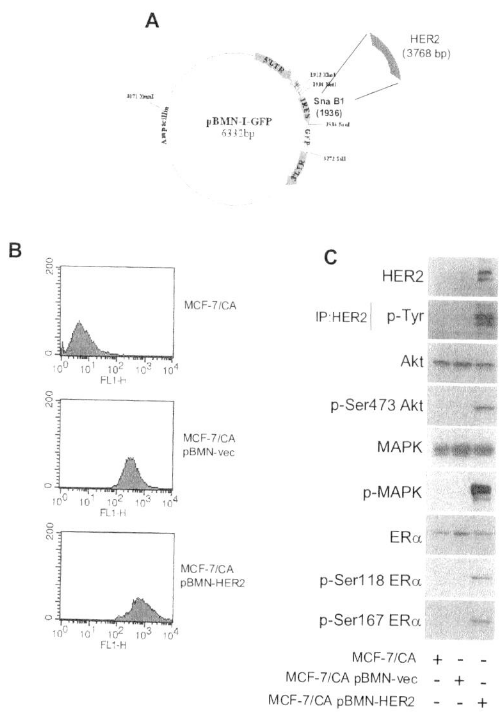

Fig. 1. HER2 overexpression in MCF-7/CA cells leads to ERα phosphorylation. A. MCF-7/CA cells infected with retrovirus generated either by pBMN-IRES-EGFP or pBMN-HER2-IRES-EGFP were sorted for EGFP expression by using flow cytometry. After few passages, the cells were analyzed with flow cytometry again to check the expression of EGFP before the experiments. As compared to parental MCF-7/CA cells, more than 98 % of both MCF-7/CA pBMN-vec and MCF-7/CA pBMN-HER2 were EGFP positive. B. Cells growing in IMEM supplemented with 10 % FBS were harvested and prepared for immunoblots for each indicated antibodies as described in MATERALS and METHODS. C. MCF-7/CA pBMN-vec cells showed increase phosphorylation of ERα on both Akt and MAPK sites.

HER2 overexpression impairs inhibitory effect of letrozole on ERE-mediated transcription and cell proliferation

Given that MCF-7/CA pBMN-HER2 cells showing high MAPK and Akt activities which are known to be associated with aberrant cell proliferation, it was a logical next step to compare MCF-7/CA pBMN-vec with MCF-7/CA pBMN-HER2 cells for ER-mediated transcription and cell proliferation. As shown in Fig. 2A, AD treatment (25 nM) for 24 h increased ER-mediated transcription by 5 fold as determined by ERE (estrogen responsive element)-luciferase reporter assays in MCF-7/CA pBMN-vec cells. The optimal concentration of AD used was 25 nM because 25 nM AD induced almost the same level of ERE-reporter activity as to the cells treated with 1 nM E2. This is in agreement with published results by Thiantanawat *et al.*(Thiantanawat et al. 2003) in which the authors showed 25 nM AD induced comparable level of response as 1 nM E2 in their MCF-7/CA cells with different origin. While letrozole could not affect E2-mediated transcription, it could completely abrogate AD-mediated transcription in MCF-7/CA pBMN-vec cells. However in MCF-7/CA pBMN-HER2 cells, we observed higher basal level of ERE-reporter activity and more than 2 fold induction of transcription over vector control cells with AD treatment. Moreover, letrozole treatment could not completely inhibit AD-mediated transcription as in MCF-7/CA pBMN-vec cells even at the concentration of 1 μM which is ten fold higher than the concentration sufficient to totally inhibit AD-mediated transcription in MCF-7/CA pBMN-vec cells. E2-mediated transcriptional activity is still higher in MCF-7/CA pBMN-HER2 cells compared with MCF-7/CA pBMN-HER2 cells, which is in good agreement with our previous report (Kurokawa et al. 2000). We next examined whether blocking HER2-induced intracellular signaling with pharmacological inhibitors would abrogate HER2-mediated increase in ERE-reporter activity. As shown in Fig.2B, an inhibitor of phosphatidylinositol-3-kinase (PI3K), LY294002 (Vlahos et al. 1994) and U0126, an inhibitor of MAP kinase kinase (MEK) (Favata et al. 1998) could partially inhibit AD-mediated ERE-reporter activity in both MCF-7/CA pBMN-vec cells and MCF-7/CA pBMN-HER2 cells. ZD1839 at the concentration of 3 μM, which was known to inhibit HER2 phosphorylation in breast cancer cells (Moulder et al. 2001), could inhibit AD-induced ERE-reporter activity to a greater extent than the other inhibitors in both cell lines. The efficacy of these inhibitors was confirmed by immunoblots shown in Fig. 2C. LY294002 could inhibit phosphorylation of Akt and Ser167 phosphorylation on ERα and U0126 could inhibit phosphorylation of MAPK and Ser118 on ERα. ZD1839 efficiently blocked tphosphorylation of tyrosine residues on HER2, Akt, MAPK and both Ser167 and Ser118 residues on ERα. Taken together the data in Fig. 2 suggest that overexpressed HER2 induced an increase in ERE-reporter activity through Akt and MAPK signaling and subsequent phosphorylation of ERα on both Ser167 and Ser118 residues. The phosphorylation of ERα on both residues was known to induce ligand-independent receptor activation and activate ERα-mediated transcription (Kato et al. 1995, Campbell et al. 2001)

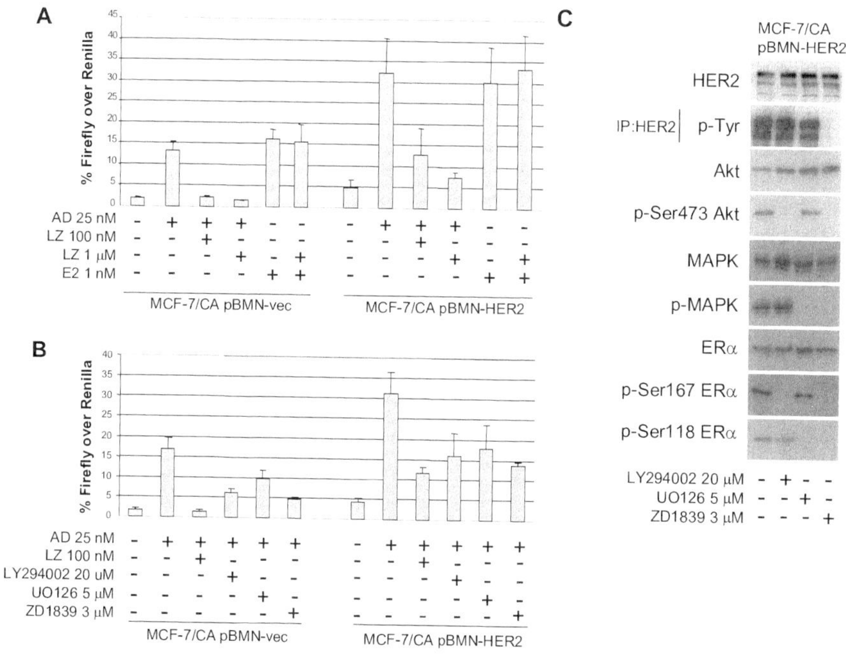

Fig. 2. Effect of HER2 overexpression on letrozole-mediated inhibition of AD-induced ERE-luciferase activity. A/B. The cells were transfected with pGLB-MERE and pCMV-Rl, treated with each indicated hormones and inhibitors and analyzed by dual luciferase assay (MATERIALS AND METHODS). Each bar represents mean±S.D. of triplicate wells repeated at least twice. C. The cells were treated with each indicated inhibitors for 24 h in IMEM supplemented with 10 % FBS, harvested and prepared for immunoblots.

We next examined whether HER2 overexpression would confer resistance to letrozole effect on cell growth. Both MCF-7/CA pBMN-vec and MCF-7/CA pBMN-HER2 cells were hormone-starved for 5 days before stimulation with 25 nM AD ± 100 nM letrozole. After 6 days of incubation the cell numbers were counted. As shown in Fig. 3A, AD treatment significantly increased cell proliferation over the control cells in a hormone-free medium. The co-treatment of letrozole could completely abrogate the growth of MCF-7/CA pBMN-vec cells but had no effect on E2-mediated growth. When MCF-7/CA pBMN-HER2 cells were treated with AD and E2, around 2-fold increase in cell proliferation was observed as compared to MCF-7/CA pBMN-vec cells.

Estrogen-deprivation induced apoptosis is inhibited by HER2 overexpression

The effect of HER2 overexpression on estrogen-deprivation induced apoptosis was next tested. Both MCF-7/CA pBMN-vec and MCF-7/CA pBMN-HER2 cells were plated in a phenol-red free medium containing 10 % of charcoal stripped serum supplemented with 25 nM AD ± 100 nM letrozole. Blockade of aromatase activity by letrozole led to estrogen deprivation in the medium, inducing apoptosis of estrogen-dependent MCF-7 cells. In 96 h after co-treatment of letrozole with AD, 8.50 % of MCF-7/CA pBMN-vec cells were appeared as apoptotic by Apo-BrdU assay (MATERIALS AND METHODS) while 3.18 % of cells were apoptotic in the absence of letrozole (Fig. 4A). In MCF-7/CA pBMN-HER2 cells, there was less basal level apoptosis than MCF-7/CA pBMN-vec cells when the cells were treated with AD alone (1.38 % to 3.18 %) and decreased apoptosis was also observed by letrozole co-treatment (2.65 % to 8.50 %). The induction of apoptosis in MCF-7/CA pBMN-vec cells was confirmed by caspase-3 cleavage, which is a hallmark of apoptosis (Nicholson et al. 1995). The cleaved caspase-3 was clearly observed only in MCF-7/CA pBMN-vec cells co-treated with AD and letrozole. These data suggest that HER2 overexpression prevented apoptosis of MCF-7/CA cells induced by estrogen deprivation and that increased survival which is partially responsible for the enhanced proliferation of MCF-7/CA pBMN-HER2 cells.

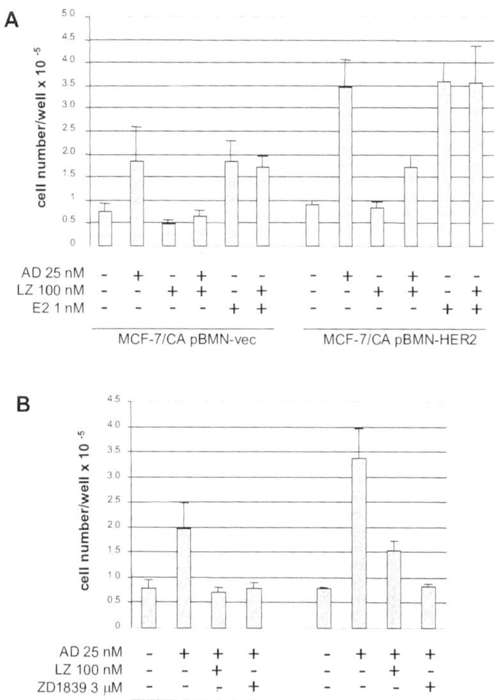

Fig. 3. HER2 overexpression abrogates letrozole-induced inhibition of cell proliferation. A/B. Cells were hormone-deprived for 3 days before seeding in 12-well plates. After seeding, the fresh hormone-deprived medium containing 25 nM AD ± 100 nM letrozole or 1 μM E2 ± 100 nM letrozole were replenished every 2 days for additional 8 days and cell numbers were counted. For the experiments in panel B, 3 μM ZS1839 was co-treated with 25 nM AD.

In contrast to the results with MCF-7/CA pBMN-vec cells, co-treatment of letrozole with AD could only partially inhibit AD-mediated proliferation of MCF-7/CA pBMN-HER2 cells. Inhibition of HER2 receptor kinase activity by ZD1839 significantly blocked AD-induced cell proliferation in both cells (Fig. 3B), suggesting that HER2 signaling confers resistance to letrozole-mediated inhibition of cell proliferation.

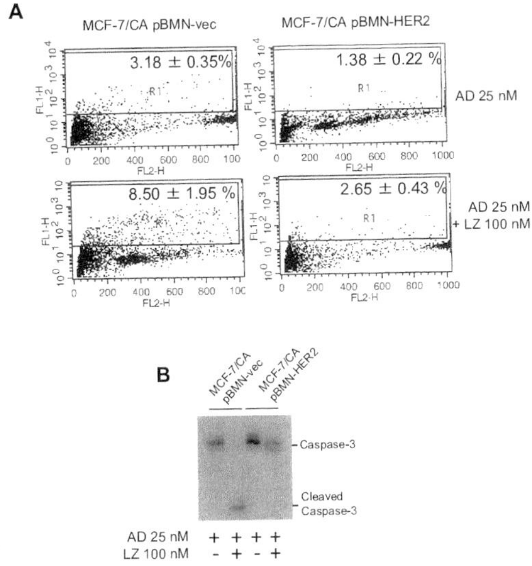

Fig. 4. Letrozole-induced apoptosis was abrogated in HER2 overexpressing cells. A. The cells were incubated in hormone-deprived medium supplemented with 25 nM AD ± 100 nM letrozol for 4 days and analyzed for TUNEL staining with Apo-BrdU assay kit. Percentage of TUNEL positive cells were depicted on each panels. B. The cells were treated as in A and analyzed for caspase-3 cleavage by immunoblot with total caspase-3 antibody.

Letrozole resistance is not mediated by increased ER-HER2 cross-talk nor resistance of aromatase activity *per se* in HER2 overexpressing cells.

Recently, Shou *et al.* (Shou et al. 2004) reported bidirectional cross-talk between ER and HER1/2 in HER2 overexpressing MCF-7 cells. In their experiment, treatment with estrogen and tamoxifen induced phosphorylation of HER1, HER2, Akt and MAPK. They proposed that growth of MCF-7/HER2 tumors is dependent on cross-talk between ER and HER2. Hence we examined whether the same parallel could be drawn in our system with AD as an agonist. However, we could not detect any modulation of HER2 phosphorylation either by E2 or AD (Fig. 5A). The phosphorylation of downstream signaling molecules like MAPK and Akt were also not affected by E2 or AD in MCF-7/CA pBMN-HER2 cells (data not shown).

To rule out the possibility that HER2 overexpression modulates intrinsic properties of aromatase enzyme to induce increased response toward AD, aromatase activities of parental MCF-7/CA, MCF-7/CA pBMN-vec and MCF-7/CA pBMN-HER2 cells were examined by tritiated water release assay (MATERIALS AND METHODS) (Fig. 5B). Although MCF-7/CA pBMN-HER2 cells showed little higher activity than other two cell lines, the aromatase activities of all three lines were efiiciently blocked by treatment with letrozole. These results indicate that resistance to letrozole iexpressed as ERE-reporter activity and cell proliferation (Figs 2 and 3) was not from the acquired resistance of aromatase enzyme to letrozole in MCF-7/CA pBMN-HER2 cells. We cannot rule out the possibility that enhanced response to AD in MCF-7/CA pBMN-HER2 cells was resulted from post-translational modification of aromatase by HER2 signaling (DISCUSSION). In accordance with previous report (Zhou et al. 1993), MCF-7 cells which was not stably transfected with aromatase (Fig. 5B, 7 and 8th bars), show very low aromatase activity, even though the aromatase gene is amplified in MCF-7 cells. The suppressive mechanism of aromatase activity is still not explained.

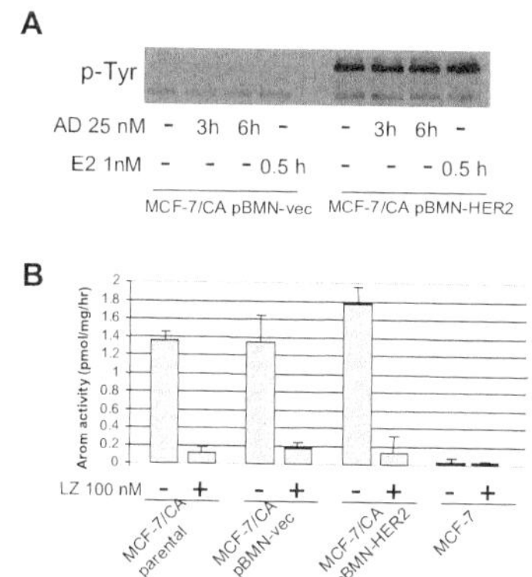

Fig. 5. Letrozole-resistance is not mediated by non-genomic ER signaling nor resistance to aromatase enzyme activity to letrozole in HER2 overexpressing cells. A. Cells were treated with 1 nM E2 or 25 nM AD for each indicated time points, harvested and analyzed with phospho-tyrosine immunoblot. B. Cells were labeled with 1 μCi/ml of [^{3}H]-AD for 3 h in phenol red-free IMEM supplemented with charcoal stripped serum and cell labeling medium was collected and analyzed for tritiated water relase assay. The protein mass of the cells were measured and aromatase activity was normalized with respect to protein mass in each wells. Each bar represents mean ± S.D. in trplicate wells.

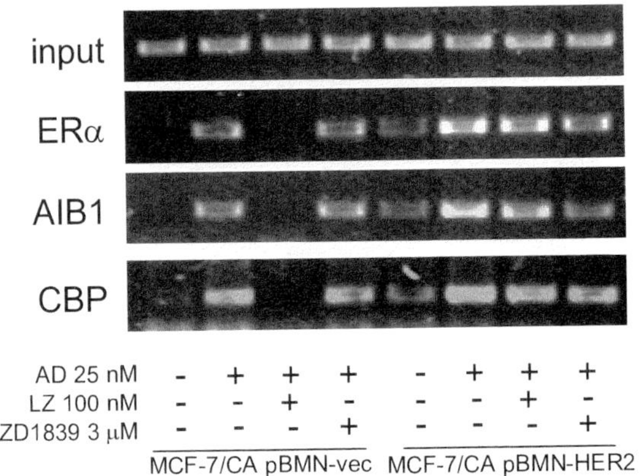

Fig. 6. HER2 overexpression recruits ERα and co-activators to pS2 promoter region in a ligand-indpendent manner and abrogates letrozole inhibition of AD-induced transcription complex recruitment. Cells were prepared for ChIP assay as described in MATERIALS AND METHODS. Association of ERα and co-activators to pS2 promoter region was determined by PCR amplification of the cross-linked DNA immunoprecipitated with each antibodies.

Increased recruitment of co-activators to ER transcription component by HER2 overexpression

Having convinced that HER2 expression confers resistance to letrozole in MCF-7/CA pBMN-HER2 cells, we next analyzed the assembly of ER transcription complex on previously characterized estrogen regulated gene pS2 (Berry et al. 1989) promoter by ChIP (chromatin immunoprecipitation) assay (Fig. 6). In MCF-7/CA pBMN-vec cells, AD treatment induced occupancy of the pS2 promoter by EvR and coactivators, amplified in breast cancer 1 (AIB1) and CREB binding protein (CBP) which interact with ER in a ligand-dependent manner (Iwase 2003). This indicates that AD is converted into estrogen by aromatase activity in these cells and interacts with ERα. Upon inhibition of aromatase by letrozole, the association of ERα and the coactivators were completely abrogated in MCF-7/CA pBMN-vec cells (Fig. 6 lane 2 vs 3). However, HER2 overexpression not only increased basal level of ligand-independent association of ERα and coactivators, but also partially blocked inhibitory effect of letrozole on AD-induced transcription complex formation in MCF-7/CA pBMN-HER2 cells (Fig. 6 lane 6 vs 7). The attenuated inhibitory effect of letrozole on transcription complex formation in MCF-7/CA pBMN-HER2 cells maybe attributable to HER2 signaling because ZD1839 treatment could also partially block AD-induced transcription complex formation in MCF-7/CA pBMN-HER2 cells (Fig. 6 lane 6 vs 8).

Discussion

We have previously described the role of HER2 overexpression on tamoxifen resistance In MCF-7 cells and that the inhibition of HER2 and MAPK restored the inhibitory effect of tamoxifen on ER-mediated transcription and cell growth (Kurokawa et al. 2000). Because chronic deprivation of estrogen led to hyperactivation of MAPK in breast cancer cells (Shim et al. 2000, Coutts and Murphy 1998), we postulated that antiproliferative effect of estrogen deprivation by aromatase inhibitor could be counteracted by elevated signaling from HER2 overexpression. To prove that hypothesis, we used MCF-7 cells stably transfected with cellular aromatase cultured in a steroid-free medium containing AD as a substrate for aromatase. Indeed, forced expression of HER2 in MCF-7/CA cells resulted in a partial resistance to letrozole treatment. AD-induced ERE-luciferase activity was resistant to letrozole in MCF-7/CA pBMN-HER2 cells and cell proliferation driven by AD in a hormone-free medium was also resistance to letrozole-induced inhibition (Figs 2 and 3).

Enhanced survival of HER2 overexpressing cells in the presence of letrozole (Fig. 4) might be responsible for the increased proliferation of these cells in an estrogen-deprived medium. As previously reported (Thiantanawat et al. 2003), letrozole induces apoptosis of MCF-7/CA cells in a hormone-free medium supplemented with AD. Interestingly, letrozole was shown to be most effective in inducing apoptosis compared with antiestrogens like tamoxifen and faslodex or estrogen deprivation and it is even more effective than the other aromatase inhibitor, anastrozole (Thiantanawat et al. 2003). Though it was reported that MCF-7 cells lack functional caspase-3 (Kurokawa et al. 1999), we could detect both pro-caspase-3 and cleaved caspase-3 by immunoblot (Fig. 4B) in our MCF-7/CA cells. In MCF-7/CA cells devoid of functional caspase-3, aromatase inhibitors and antiestrogens could induce apoptosis by down-regulation of Bcl-2, upregulation of Bax, and caspase-6,7,9 activation (Thiantanawat et al. 2003). Since both caspase-3 and caspase-7 act downstream of caspase-9 (Fraser and Evan 1996) and share similar substrate specificity (i.e. poly-ADP-ribose polymerase) (Thornberry et al. 1997), the downstream machinery of apoptosis may be shared by both caspase-3 deficient cells and caspase-3 positive cells.

As briefly mentioned in the RESULTS section, Shou *et al.* (Shou et al. 2004) reported bidirectional molecular cross-talk between HER2 signaling and ER signaling in their MCF-7 overexpressing HER2 (MCF-7/HER2-18) clones. In their experiments, both estrogen and tamoxifen treatment phosphorylated HER1, HER2, Akt and MAPK to the extent similar to the levels induced by EGF and heregulin treatment in MCF-7/HER2-18 cells. They also showed that ZD1839 minimally affected MCF-7/HER2-18 tumor growth in mice treated with estrogen while totally blocking tamoxifen-stimulated growth. Their data strongly support a rationale for combining tamoxifen with ZD1839 for the treatment of ER-positive HER2-overexpressing breast cancer. Unexpectedly, we could not observe AD- or E2-induced phosphorylation of HER2 or downstream signaling molecules in our MCF-7/CA pBMN-HER2 cells (Fig. 5A and data not shown). This discrepancy might be explained by the differences in HER2 expression level in both cell lines, although we could not completely rule out the possibility that stable expression of aromatase might somehow interfere with cross-talk between ER and HER family receptor tyrosine kinases.

ER was reported to bind to p85 regulatory subunit of PI3K in a ligand dependent manner, leading to the activation of Akt and endothelial nitric oxide synthase in human endothelial cells (Simoncini et al. 2000), implicating another non-nuclear signaling pathway mediated by ER. Non-genomic ERα signaling was also implicated in Src/Ras/MAPK pathway (Migliaccio et al. 1996, Migliaccio et al. 1998, Improta-Brears et al. 1999)

Regulation of aromatase activity is mainly achieved by transcriptional regulation in a tissue specific manner (Sasano and Harada 1998).The promoters in exon I.3 and II are the major promoters driving aromatase expression in breat cancer cells (Zhou et al. 1996, Utsumi et al. 1996). Expression of aromatase transcripts via those promoters is regulated by prostaglandin E2 (Zhao et al. 1996) via cAMP (Mahendroo et al. 1993). Conversely, cAMP inhibits dexamethasone and phorbol ester-induced aromatase activity in THP-1 cells (Shozu et al. 1997). Recently, post-translational regulation of aromatase has been suggested. Shozu *et al.* (Shozu et al. 2001) phosphorylation. Though their study mainly relied on the use of pharmacological inhibitors of MEK, their results, together with another report showing that the aromatase activity is more stable in the presence of the phosphatase inhibitors (Bellino and Holben 1989), it could be postulated that aromatase activity might be up-regulated by protein phosphorylation. However, recent reports by another group suggest an opposite regulatory mechanism by phosphorylation. Balthazart *et al* (Balthazart et al. 2001, Balthazart et al. 2003) showed that aromatase activity in quail hypothalamic homogenates is down-regulated by calcium-dependent phosphorylation and this inhibition is blocked by kinase inhibitors. Recently, it was reported that E2 induces aromatase activity both in SK-BR-3 and MCF-7 cells when ERα is ectopially overexpressed (Kinoshita and Chen 2003). In their report, E2-induced aromatase activity in HER2-overexpressing MCF-7 cells after transfection with ERα is several folds higher than that found in normal MCF-7 cells, supporting a cross-talk mechanism between ERα and HER-2. Another supporting data is that the increase in aromatase activity by E2 in ERα-transfected cells is effectively inhibited by MEK inhibitor, PD98059 and HER1 tyrosine kinase inhibitor, PD153035 in their experiments.

In Fig. 5B, the moderate increase of aromatase activity in MCF-7/CA pBMN-HER2 cells over vector cells might be resulted from upregulation of aromatase activity by HER2 mediated signaling. Indeed, we observed that a decrease in aromatase activity in MCF-7/CA cells treated with inhibitors of PI3K and MAPK signaling and co-expression of constitutive active Akt with aromatase expression vector in Cos-7 cells leading to an increase in aromatase activity [1]. Further experiments are currently undergoing to confirm the regulation of aromatase by phosphorylation. Though we observed a slight increase of aromatase activity in HER2 overexpressing cells, treatment with 100 nM letrozole could still abrogate aromatase activity in these cells. This suggests that resistance to letrozole was not mediated by the acquired resistance of aromatase enzyme by increased HER2 signaling.

Upon binding of E2, ER activates transcription through association with ERE located within the promoter regions of targent genes (Robinson-Rechavi et al. 2003). ERα undergoes structural rearrangements after association with ligands to expose binding surfaces to recruit transcription cofactors (Brzozowski et al. 1997). pS2 has been well characterized as an E2-inducible gene in MCF-7 cells (Masiakowski et al. 1982) and in MCF-7 cells, E2 treatment recruits ERα and coactivators AIB1 and histone acetylases P300 and

---

[1] Shin *et al.* unpublished data

CBP to pS2 gene promoter region (Shou et al. 2004). As shown in Fig. 6, HER2 overexpressing cells exhibited higher basal level association of ERα and coactivators AIB1 and CBP to pS2 promoter region and AD-induced association of transcription complex formation was partially inhibited by letrozole and ZD1839 treatment, suggesting that the enhanced HER2 signaling could mediate letrozole resistance at least in part by maintaining constitutive association of transcription complex. The HER2-mediated increase in transcription complex association was also observed in MCF-7/HER2-18 cells (Shou et al. 2004). In their experiment, both E2 and tamoxifen-induced recruitments of ERα, AIB1 p300 and CBP are enhanced in HER2-overexpressing cells and ZD1839 treatment decreases the transcription complex associations to the levels comparable to parental MCF-7 cells, while inducing recruitment of co-repressors NCoR and HDAC3 by tamoxifen in MCF-7/HER2-18 cells.

Recently, Ellis *et al.* (Ellis et al. 2001, Ellis et al. 2003) suggested that letrozole inhibited tumor proliferation to a greater extent than tamoxifen in ER positive, HER2 and/or HER1 positive tumors, implicating that tumors positive for HER2 or HER1 are highly dependent on ER signaling. However, our data indicate that significant levels of ERα and coactivators on the pS2 promoter in the absence of hormones in MCF-7/CA pBMN-HER2 cells (Fig. 6), suggesting that overexpression of HER2 could partially overcome estrogen deprivation mediated by letrozole treatment. In summary, our data propose that the combined treatment of aromatase inhibitors with HER2 signaling inhibitors might be beneficial to the treatment of ER and HER2 positive breast tumors.

References

Andrulis IL, Bull SB, Blackstein ME, Sutherland D, Mak C, Sidlofsky S, Pritzker KP, Hartwick RW, Hanna W, Lickley L, Wilkinson R, Qizilbash A, Ambus U, Lipa M, Weizel H, Katz A, Baida M, Mariz S, Stoik G, Dacamara P, Strongitharm D Geddie W,McCready D (1998) neu/erbB-2 amplification identifies a poor-prognosis group of women with node-negative breast cancer. Toronto Breast Cancer Study Group. J Clin Oncol 16: 1340-1349

Balthazart J, Baillien M, Charlier TD,Ball GF (2003) Calcium-dependent phosphorylation processes control brain aromatase in quail. Eur J Neurosci 17: 1591-1606

Balthazart J, Baillien M,Ball GF (2001) Phosphorylation processes mediate rapid changes of brain aromatase activity. J Steroid Biochem Mol Biol 79: 261-277

Bellino FL,Holben L (1989) Placental estrogen synthetase (aromatase): evidence for phosphatase-dependent inactivation. Biochem Biophys Res Commun 162: 498-504

Benz CC, Scott GK, Sarup JC, Johnson RM, Tripathy D, Coronado E, Shepard HM,Osborne CK (1992) Estrogen-dependent, tamoxifen-resistant tumorigenic growth of MCF-7 cells transfected with HER2/neu. Breast Cancer Res Treat 24: 85-95

Berry M, Nunez AM,Chambon P (1989) Estrogen-responsive element of the human pS2 gene is an imperfectly palindromic sequence. Proc Natl Acad Sci U S A 86: 1218-1222

Brzozowski AM, Pike AC, Dauter Z, Hubbard RE, Bonn T, Engstrom O, Ohman L, Greene GL, Gustafsson JA,Carlquist M (1997) Molecular basis of agonism and antagonism in the oestrogen receptor. Nature 389: 753-758

Bulun SE, Noble LS, Takayama K, Michael MD, Agarwal V, Fisher C, Zhao Y, Hinshelwood MM, Ito Y,Simpson ER (1997) Endocrine disorders associated with inappropriately high aromatase expression. J Steroid Biochem Mol Biol 61: 133-139

Bulun SE, Price TM, Aitken J, Mahendroo MS,Simpson ER (1993) A link between breast cancer and local estrogen biosynthesis suggested by quantification of breast adipose tissue aromatase cytochrome P450 transcripts using competitive polymerase chain reaction after reverse transcription. J Clin Endocrinol Metab 77: 1622-1628

Buzdar AU, Jonat W, Howell A, Jones SE, Blomqvist CP, Vogel CL, Eiermann W, Wolter JM, Steinberg M, Webster A,Lee D (1998) Anastrozole versus megestrol acetate in the treatment of postmenopausal women with advanced breast carcinoma: results of a survival update based on a combined analysis of data from two mature phase III trials. Arimidex Study Group. Cancer 83: 1142-1152

Campbell RA, Bhat-Nakshatri P, Patel NM, Constantinidou D, Ali S,Nakshatri H (2001) Phosphatidylinositol 3-kinase/AKT-mediated activation of estrogen receptor alpha: a new model for anti-estrogen resistance. J Biol Chem 276: 9817-9824

Coutts AS,Murphy LC (1998) Elevated mitogen-activated protein kinase activity in estrogen-nonresponsive human breast cancer cells. Cancer Res 58: 4071-4074

Dombernowsky P, Smith I, Falkson G, Leonard R, Panasci L, Bellmunt J, Bezwoda W, Gardin G, Gudgeon A, Morgan M, Fornasiero A, Hoffmann W, Michel J, Hatschek T, Tjabbes T, Chaudri HA, Hornberger U,Trunet PF (1998) Letrozole, a new oral aromatase inhibitor for advanced breast cancer: double-blind randomized trial showing a dose effect and improved efficacy and tolerability compared with megestrol acetate. J Clin Oncol 16: 453-461

Ellis MJ, Coop A, Singh B, Mauriac L, Llombert-Cussac A, Janicke F, Miller WR, Evans DB, Dugan M, Brady C, Quebe-Fehling E,Borgs M (2001) Letrozole is more effective neoadjuvant endocrine therapy than tamoxifen for ErbB-1- and/or ErbB-2-positive, estrogen receptor-positive primary breast cancer: evidence from a phase III randomized trial. J Clin Oncol 19: 3808-3816

Ellis MJ, Coop A, Singh B, Tao Y, Llombart-Cussac A, Janicke F, Mauriac L, Quebe-Fehling E, Chaudri-Ross HA, Evans DB,Miller WR (2003) Letrozole inhibits tumor proliferation more effectively than tamoxifen independent of HER1/2 expression status. Cancer Res 63: 6523-6531

Favata MF, Horiuchi KY, Manos EJ, Daulerio AJ, Stradley DA, Feeser WS, Van Dyk DE, Pitts WJ, Earl RA, Hobbs F, Copeland RA, Magolda RL, Scherle PA,Trzaskos JM (1998) Identification of a novel inhibitor of mitogen-activated protein kinase kinase. J Biol Chem 273: 18623-18632

Fraser A,Evan G (1996) A license to kill. Cell 85: 781-784

Graus-Porta D, Beerli RR, Daly JM,Hynes NE (1997) ErbB-2, the preferred heterodimerization partner of all ErbB receptors, is a mediator of lateral signaling. Embo J 16: 1647-1655

Hemsell DL, Grodin JM, Brenner PF, Siiteri PK,MacDonald PC (1974) Plasma precursors of estrogen. II. Correlation of the extent of conversion of plasma androstenedione to estrone with age. J Clin Endocrinol Metab 38: 476-479

Houston SJ, Plunkett TA, Barnes DM, Smith P, Rubens RD,Miles DW (1999) Overexpression of c-erbB2 is an independent marker of resistance to endocrine therapy in advanced breast cancer. Br J Cancer 79: 1220-1226

Improta-Brears T, Whorton AR, Codazzi F, York JD, Meyer T,McDonnell DP (1999) Estrogen-induced activation of mitogen-activated protein kinase requires mobilization of intracellular calcium. Proc Natl Acad Sci U S A 96: 4686-4691

Iwase H (2003) Molecular action of the estrogen receptor and hormone dependency in breast cancer. Breast Cancer 10: 89-96

Janes PW, Daly RJ, deFazio A,Sutherland RL (1994) Activation of the Ras signalling pathway in human breast cancer cells overexpressing erbB-2. Oncogene 9: 3601-3608

Kato S, Endoh H, Masuhiro Y, Kitamoto T, Uchiyama S, Sasaki H, Masushige S, Gotoh Y, Nishida E, Kawashima H, Metzger D,Chambon P (1995) Activation of the estrogen receptor through phosphorylation by mitogen-activated protein kinase. Science 270: 1491-1494

Kaufmann M, Bajetta E, Dirix LY, Fein LE, Jones SE, Zilembo N, Dugardyn JL, Nasurdi C, Mennel RG, Cervek J, Fowst C, Polli A, di Salle E, Arkhipov A, Piscitelli G, Miller LL,Massimini G (2000) Exemestane improves survival in metastatic breast cancer: results of a phase III randomized study. Clin Breast Cancer 1 Suppl 1: S15-18

Kinoshita Y,Chen S (2003) Induction of aromatase (CYP19) expression in breast cancer cells through a nongenomic action of estrogen receptor alpha. Cancer Res 63: 3546-3555

Kurokawa H, Lenferink AE, Simpson JF, Pisacane PI, Sliwkowski MX, Forbes JT,Arteaga CL (2000) Inhibition of HER2/neu (erbB-2) and mitogen-activated protein kinases enhances tamoxifen action against HER2-overexpressing, tamoxifen-resistant breast cancer cells. Cancer Res 60: 5887-5894

Kurokawa H, Nishio K, Fukumoto H, Tomonari A, Suzuki T,Saijo N (1999) Alteration of caspase-3 (CPP32/Yama/apopain) in wild-type MCF-7, breast cancer cells. Oncol Rep 6: 33-37

Leonessa F, Boulay V, Wright A, Thompson EW, Brunner N,Clarke R (1992) The biology of breast tumor progression. Acquisition of hormone independence and resistance to cytotoxic drugs. Acta Oncol 31: 115-123

Litherland S,Jackson IM (1988) Antioestrogens in the management of hormone-dependent cancer. Cancer Treat Rev 15: 183-194

Liu Y, el-Ashry D, Chen D, Ding IY,Kern FG (1995) MCF-7 breast cancer cells overexpressing transfected c-erbB-2 have an in vitro growth advantage in estrogen-depleted conditions and reduced estrogen-dependence and tamoxifen-sensitivity in vivo. Breast Cancer Res Treat 34: 97-117

Long BJ, Jelovac D, Thiantanawat A,Brodie AM (2002) The effect of second-line antiestrogen therapy on breast tumor growth after first-line treatment with the aromatase inhibitor letrozole: long-term studies using the intratumoral aromatase postmenopausal breast cancer model. Clin Cancer Res 8: 2378-2388

Long BJ, Tilghman SL, Yue W, Thiantanawat A, Grigoryev DN,Brodie AM (1998) The steroidal antiestrogen ICI 182,780 is an inhibitor of cellular aromatase activity. J Steroid Biochem Mol Biol 67: 293-304

Mahendroo MS, Mendelson CR,Simpson ER (1993) Tissue-specific and hormonally controlled alternative promoters regulate aromatase cytochrome P450 gene expression in human adipose tissue. J Biol Chem 268: 19463-19470

Masiakowski P, Breathnach R, Bloch J, Gannon F, Krust A,Chambon P (1982) Cloning of cDNA sequences of hormone-regulated genes from the MCF-7 human breast cancer cell line. Nucleic Acids Res 10: 7895-7903

Means GD, Kilgore MW, Mahendroo MS, Mendelson CR,Simpson ER (1991) Tissue-specific promoters regulate aromatase cytochrome P450 gene expression in human ovary and fetal tissues. Mol Endocrinol 5: 2005-2013

Migliaccio A, Di Domenico M, Castoria G, de Falco A, Bontempo P, Nola E,Auricchio F (1996) Tyrosine kinase/p21ras/MAP-kinase pathway activation by estradiol-receptor complex in MCF-7 cells. Embo J 15: 1292-1300

Migliaccio A, Piccolo D, Castoria G, Di Domenico M, Bilancio A, Lombardi M, Gong W, Beato M,Auricchio F (1998) Activation of the Src/p21ras/Erk pathway by progesterone receptor via cross-talk with estrogen receptor. Embo J 17: 2008-2018

Moulder SL, Yakes FM, Muthuswamy SK, Bianco R, Simpson JF,Arteaga CL (2001) Epidermal growth factor receptor (HER1) tyrosine kinase inhibitor ZD1839 (Iressa) inhibits HER2/neu (erbB2)-overexpressing breast cancer cells in vitro and in vivo. Cancer Res 61: 8887-8895

Mouridsen H, Gershanovich M, Sun Y, Perez-Carrion R, Boni C, Monnier A, Apffelstaedt J, Smith R, Sleeboom HP, Janicke F, Pluzanska A, Dank M, Becquart D, Bapsy PP, Salminen E, Snyder R, Lassus M, Verbeek JA, Staffler B, Chaudri-Ross HA,Dugan M (2001) Superior efficacy of letrozole versus tamoxifen as first-line therapy for postmenopausal women with advanced breast cancer: results of a phase III study of the International Letrozole Breast Cancer Group. J Clin Oncol 19: 2596-2606

Naftolin F, Ryan KJ, Davies IJ, Reddy VV, Flores F, Petro Z, Kuhn M, White RJ, Takaoka Y,Wolin L (1975) The formation of estrogens by central neuroendocrine tissues. Recent Prog Horm Res 31: 295-319

Nicholson DW, Ali A, Thornberry NA, Vaillancourt JP, Ding CK, Gallant M, Gareau Y, Griffin PR, Labelle M, Lazebnik YA,et al. (1995) Identification and inhibition of the ICE/CED-3 protease necessary for mammalian apoptosis. Nature 376: 37-43

Olayioye MA, Neve RM, Lane HA,Hynes NE (2000) The ErbB signaling network: receptor heterodimerization in development and cancer. Embo J 19: 3159-3167

Perel E, Blackstein ME,Killinger DW (1982) Aromatase in human breast carcinoma. Cancer Res 42: 3369s-3372s

Riese DJ, 2nd,Stern DF (1998) Specificity within the EGF family/ErbB receptor family signaling network. Bioessays 20: 41-48

Robinson-Rechavi M, Escriva Garcia H,Laudet V (2003) The nuclear receptor superfamily. J Cell Sci 116: 585-586

Sasano H,Harada N (1998) Intratumoral aromatase in human breast, endometrial, and ovarian malignancies. Endocr Rev 19: 593-607

Seralini G,Moslemi S (2001) Aromatase inhibitors: past, present and future. Mol Cell Endocrinol 178: 117-131

Shim WS, Conaway M, Masamura S, Yue W, Wang JP, Kmar R,Santen RJ (2000) Estradiol hypersensitivity and mitogen-activated protein kinase expression in long-term estrogen deprived human breast cancer cells in vivo. Endocrinology 141: 396-405

Shou J, Massarweh S, Osborne CK, Wakeling AE, Ali S, Weiss H,Schiff R (2004) Mechanisms of tamoxifen resistance: increased estrogen receptor-HER2/neu cross-talk in ER/HER2-positive breast cancer. J Natl Cancer Inst 96: 926-935

Shozu M, Sumitani H, Murakami K, Segawa T, Yang HJ,Inoue M (2001) Regulation of aromatase activity in bone-derived cells: possible role of mitogen-activated protein kinase. J Steroid Biochem Mol Biol 79: 61-65

Shozu M, Zhao Y,Simpson ER (1997) Estrogen biosynthesis in THP1 cells is regulated by promoter switching of the aromatase (CYP19) gene. Endocrinology 138: 5125-5135

Shozu M,Simpson ER (1998) Aromatase expression of human osteoblast-like cells. Mol Cell Endocrinol 139: 117-129

Silva MC, Rowlands MG, Dowsett M, Gusterson B, McKinna JA, Fryatt I,Coombes RC (1989) Intratumoral aromatase as a prognostic factor in human breast carcinoma. Cancer Res 49: 2588-2591

Simoncini T, Hafezi-Moghadam A, Brazil DP, Ley K, Chin WW,Liao JK (2000) Interaction of oestrogen receptor with the regulatory subunit of phosphatidylinositol-3-OH kinase. Nature 407: 538-541

Slamon DJ, Godolphin W, Jones LA, Holt JA, Wong SG, Keith DE, Levin WJ, Stuart SG, Udove J, Ullrich A,et al. (1989) Studies of the HER-2/neu proto-oncogene in human breast and ovarian cancer. Science 244: 707-712

Thiantanawat A, Long BJ,Brodie AM (2003) Signaling pathways of apoptosis activated by aromatase inhibitors and antiestrogens. Cancer Res 63: 8037-8050

Thornberry NA, Rano TA, Peterson EP, Rasper DM, Timkey T, Garcia-Calvo M, Houtzager VM, Nordstrom PA, Roy S, Vaillancourt JP, Chapman KT,Nicholson DW (1997) A combinatorial approach defines specificities of members of the caspase family and granzyme B. Functional relationships established for key mediators of apoptosis. J Biol Chem 272: 17907-17911

Tzahar E, Pinkas-Kramarski R, Moyer JD, Klapper LN, Alroy I, Levkowitz G, Shelly M, Henis S, Eisenstein M, Ratzkin BJ, Sela M, Andrews GC,Yarden Y (1997) Bivalence of EGF-like ligands drives the ErbB signaling network. Embo J 16: 4938-4950

Tzahar E,Yarden Y (1998) The ErbB-2/HER2 oncogenic receptor of adenocarcinomas: from orphanhood to multiple stromal ligands. Biochim Biophys Acta 1377: M25-37

Ueda Y, Wang S, Dumont N, Yi JY, Koh Y,Arteaga CL (2004) Overexpression of HER2 (erbB2) in human breast epithelial cells unmasks transforming growth factor beta-induced cell motility. J Biol Chem 279: 24505-24513

Utsumi T, Harada N, Maruta M,Takagi Y (1996) Presence of alternatively spliced transcripts of aromatase gene in human breast cancer. J Clin Endocrinol Metab 81: 2344-2349

Vlahos CJ, Matter WF, Hui KY,Brown RF (1994) A specific inhibitor of phosphatidylinositol 3-kinase, 2-(4-morpholinyl)-8-phenyl-4H-1-benzopyran-4-one (LY294002). J Biol Chem 269: 5241-5248

Yamauchi H, O'Neill A, Gelman R, Carney W, Tenney DY, Hosch S,Hayes DF (1997) Prediction of response to antiestrogen therapy in advanced breast cancer patients by pretreatment circulating levels of extracellular domain of the HER-2/c-neu protein. J Clin Oncol 15: 2518-2525

Yue W, Wang JP, Conaway MR, Li Y,Santen RJ (2003) Adaptive hypersensitivity following long-term estrogen deprivation: involvement of multiple signal pathways. J Steroid Biochem Mol Biol 86: 265-274

Zhao Y, Agarwal VR, Mendelson CR,Simpson ER (1996) Estrogen biosynthesis proximal to a breast tumor is stimulated by PGE2 via cyclic AMP, leading to activation of promoter II of the CYP19 (aromatase) gene. Endocrinology 137: 5739-5742

Zhou C, Zhou D, Esteban J, Murai J, Siiteri PK, Wilczynski S,Chen S (1996) Aromatase gene expression and its exon I usage in human breast tumors. Detection of aromatase messenger RNA by reverse transcription-polymerase chain reaction. J Steroid Biochem Mol Biol 59: 163-171

Zhou D, Wang J, Chen E, Murai J, Siiteri PK,Chen S (1993) Aromatase gene is amplified in MCF-7 human breast cancer cells. J Steroid Biochem Mol Biol 46: 147-153

Biomarker Research for Development of Molecular Target Agents

Kazuto Nishio and Tokuzo Arao

Department of Genome Biology, Kinki University School of Medicine, Osakasayama, Osaka 589-8511, Japan

Summary. Recently, molecular targeting agents occupy a prominent position in the anti-cancer drug market. Many of the tyrosine kinase inhibitors have been approved for treatment of several tumor types ranging from infrequently occurring tumors to the more common lung and colon cancers. Recently, multi-tyrosine kinase inhibitors such as sorafenib and sunitunib have shown to be active against renal cell and hapatocellular carcinoma. Thus, multi-target tyrosine kinase inhibitors are highlighted. Biomarkers can adequately select a sub-population of patients designated for the specific molecular targeting agents. A clinical study of the patient sub-population can improve the efficiency of the clinical development of the anti-cancer agents. A pharmacodynamic study that monitors the targets or the surrogate markers during treatments is important and leads to the POC of the drugs. We hope to discuss following points; 1) Should a sub-population be selected by clinical factor or biomarkers? 2) How to select a standard assay system? 3) How to access the clinical samples? 4) Can circulating samples improve feasibility?

Key word. pharmacogenomics, DNA chip, biomarker, prediction

Introduction

Remarkable progress has been made in the field of molecular biology in the 20th century (Table 1). The entire human genome has been sequenced by the Human Genome Project. The 21st century is, therefore, called the "Post Genome" era and further advances in the clinical application of biotechnology are expected. Applied biotechnology is also useful for both diagnostic and therapeutic oncology. Here, we shall discuss the application of biotechnology to the field of medical oncology.

Table 1. Progress in the field of molecular biology during the 20th century

year	event
1890	Mendelism
1926	Genes on chromosome (Mogan)
1944	DNA as gene component (Eibree)
1953	Double helix of DNA (Watson & Crick)
1956	Replication enzyme of DNA (Komburg)
1973	Recombination technology (Cohen)
1985	PCR (Mullis)
1990	Start the Human Genome Project
1998	Deciphering the human genome proceed to multicellular organism
2001	Decoding of the human genome by Celera Genomics Co.

Tissue Banking

Genome biology is expected to be applied to drug development. Drug development, such as that of cytotoxic anticancer drugs and molecular target drugs in the field of oncology, is one of the most upcoming fields. The first and most important step of drug screening is target identification and the search for seeds. The next step is screening of the compounds, followed by preclinical and clinical studies. It is considered that genomic information effectively contributes to the target identification and its validation. To obtain data about the human genome, analysis of human materials is essential. This approach is called the "Reverse Translational Research". In the clinical setting, it is also called "Molecular Correlative Study". These approaches are adopted by government-supported projects both in Japan and abroad. Pharmaceutical companies also aggressively conduct a search for seeds. Mega-pharmas, in particular, have already established the banking system for human materials. Japan has also started a banking system, but it seems to be still immature and Japan still falls behind other countries. The process of collecting clinical samples is called "Tissue Banking" or simply "Banking".

Pharmacogenomics

The approach mentioned above is also applied in the clinical setting. One of the well-recognized approaches is "Personalized Medicine," that allows therapy to be customized to individuals by analyzing the individual's genome. Analysis of the genome is called "pharmacogenomics" when it is related to treatment with drugs. "Pharmacogenomics" is a word combining "genomics" and "pharmacology". Broadly, pharamcogenomics includes the analysis of gene products, such as RNA and proteins. The pharmacogenomic approach is considered to contribute to health and welfare. The US and other governments are encouraging this strategy. For example, the US government provides guidance to the industry on the process of Investigational New Drug (IND), New Drug Application (NDA), and Biologic Licence Application (BLA). In our country, the Ministry of Health, Welfare, and Labour has requested for genomic information obtained by the genomic testing in clinical studies for pharmaceutical companies.

Application of pharamcogenomics is expected in three major stages: discovery, preclinical, and clinical stages (**Table 2**). Three examples are provided as follows; i) research on gene-related diseases; ii) relationship between gene polymorphism and response to drug treatment; iii) genomic tests for the prediction of drug responses. Examples 2 and 3 are considered to be closely associated with cancer treatment and will directly contribute to the exclusion of patients with severe toxicities or to the selection of responders and non-responders to a particular treatment. The markers obtained by pharmacogenomics are called as "biomarkers".

Table 2. Three broad applications of pharmacogenomics

Discovery	
	Target identification
	Mechanisms of Action
	Target differentiation
	Biomarker identification
Preclinical Toxicology	
	Toxicogenomics
	In vivo mechanism of action
	Biomarker identification
Clinical	
	In vivo mechanism of action
	Biomarker development and validation

Biomarkers for Molecule-Targeting Target Drugs

We would like to consider biomarkers for target-based drugs. 1) Overexpression of the target molecule ; this is often detected by immunohitochemical analysis. Amplification of target molecules is detected by FISH, CISH or PCR. Somatic mutations in tumor tissues are detected by direct sequencing or other PCR- based assays. For the purification of tumor tissues, the microdissection technique is useful. There are biomarkers for conventional cytotoxic drugs. ERCC1 is an enzyme involved in DNA repair and its transcript levels have been reported to be related to the responses to platinum-containing regimens (e.g., cisplatin plus gemicitabine) in non-small cell lung cancer patients (Olaussen et al. 2006). Thus, biomarkers could be determinants for predicting the sensitivity and responses of tumors to cytotoxic drugs.

As mentioned above, the EGFR somatic mutation in lung cancer is a hot topic. Strong correlation has been observed between EGFR somatic mutations and clinical responses to an EGFR-specific tyrosine kinase inhibitor, gefitinib. Thus, the EGFR mutation is a definite biomarker, and other somatic mutations of oncogenes in tumors have been also reported. These mutations could be used as new biomarkers to clarify subpopulations of patients that would respond to molecule-targeting drugs. Currently, trials for new molecule-targeting therapeutics are now underway for solid tumors. Treatment with angiogenesis inhibitors and antibodies are expected to improve the outcome of patients. New biomarkers need to be continually sought for this type of therapeutics. Now, these molecular correlative studies are called as "Critical Path Research" in the field of drug development. Considering the background of aggressiveness of biomarker research, the average response to drugs is much lower than that of other diseases.

The average response rate to anticancer drugs is 20-30%, which is inadequate. In order to improve the response rate to anticancer drugs, selection of subpopulations of patients that would potentially show response is one strategy. At the same time, the labeling of drugs with pharmacogenomic data has been increasing recently.

Government-related regulatory institutions in the US (Department of Health and Human Services, Food and Drug Administration, Center for Drug Evaluation and Research (CDER), Center for Biologic Evaluation and Research (CBER), Center for Devices and Radiological Health (CDRH)) developed a "Guideline for Industry," by which pharmaceutical companies are required to submit pharmacogenomic data. How should investigators assess/evaluate the data? Essentially, we should recognize three categories of pharmacogenomic information while selecting the treatment strategy: 1) test required, 2) test recommended, 3) information only.

Trastuzumab (Herceptin®) for breast cancer is a good example of the first; testing for anti-Her2 by FISH analysis (Herceptest®) is required for the administration of Trastuzumab. Although EGFR somatic mutation, EGFR immunohistochemistry, and FISH for *EGFR* are considered to be good biomarkers for predicting the response to EGFR-targeting drugs, they belong to the "Test only" category. It is not within the scope of this review to discuss why these differences exist. Anyway, applied pharmacogenomics is very important in the selection of appropriate subpopulations, and an increase in the number of "Test required" biomarkers is warranted.

Another point for discussion is that the pharmacogenomic approach has so far focused on the prediction or evaluation of adverse events. Single-nucleotide polymorphisms of metabolizing enzymes, such as p450 or UDP-glucuronoyltransferases (UGT) (Innocenti et al. 2004) are closely related to the toxicity profile of drugs. Therefore, tests for these genes are also included in the label of the drugs. The available evidence actually contributes to identify subpopulations of patients likely to show severe side effects. On the other hand, there is not much evidence, in terms of biomarkers, to distinguish accurately between responders and non-responders. It is important to consider the latter approach when considering personalized medicine.

Drug-diagnostic Co-development

As mentioned before, the importance of pharmacogenomics has been discussed worldwide. Last year, the FDA proposed the new concept "drug-diagnostic co-development", although it is still in the draft stage and needs open discussion. What is the "co-development"? "Co-development" means: 1) Critical Path Research for biomarkers that would distinguish responders from non-responders in clinical studies; 2) research for avoiding severe toxicities; 3) clinical studies for POC (proof of concept) by monitoring pharmacodynamic markers. The endpoints of these approaches are to set the appropriate doses for each subpopulation or responders. Investigators should consider the study designs flexibly in these approaches. For example, randomized phase II studies and randomized discontinuation studies may be given more consideration. In addition, for the selection of biomarkers in Critical Path Research, more strict validation will be necessary, because the tests using the biomarkers will directly affect the treatment of each patient.

Problems in Pharmacogenomics and Future Perspectives

Biomarker researches can be divided into two categories, "hypothesis-driven" and "hypothesis-free"; the former is to prove the power of preexisting biomarkers (predictability, reliability, specificity *e.g.)*, whereas the latter is to select biomarkers without any hypothesis, by DNA microarray or proteomics. At the same time, validation of the selected biomarkers is necessary. Currently, the hypothesis-free approach seems to be the trend.

In general, biomarkers in the hypothesis-driven approach are relatively easy to understand, and are based on biological evidence. They can be expected to be more easily applied clinically. However, there is a limitation: only pre-existing biomarkers can be used. On the other hand, in the case of biomarkers in the hypothesis-free approach, it is difficult to understand underlying biological mechanisms and it is difficult to directly apply these markers clinically. However, novel biomarkers can be discovered by this approach.

When considering a new prospective study using microarray gene expression profiling, it is of importance to pay attention to some points, as follows. The investigators should recognize the role of quality assurance and perform the study accordingly. Regarding the data of DNA expression for Cancer Diagnostics, the guidelines proposed by the NCI-EORTC Working Group are helpful (McShane et al. 2005). For the development of classifications based on the gene expression profile, the following points must be taken into consideration. 1) A common therapy is essential for identical populations. Are the results reasonable enough to establish a therapeutic policy? Will the new classification be generally used based on the cost-benefit balance, by comparing the selection of the therapies and the cost for mis-classified? These points should be discussed preliminarily during the process of designing of the study. For further evaluation, internal validation is necessary to prove the accuracy of the new classification in comparison with the pre-existing prognostic factors. The validation process includes 1) transfer to other platforms that are commonly used in clinical situations. (For example, will the classification identified by DNA chip analysis be valid for transfer to that by RT-PCR or immunohistochemical (ICH) examination), 2) confirmation of the reproducibility of the classification by the new platform (RT-PCR or ICH), and 3) independent validation in a prospective study. In addition, the investigators should recognize "multiplicity" of the comprehensive data sets, such as those of gene expression. Many researchers have reported classifiers to predict the prognosis of patients with cancers. For example, a 17- gene signature associated with metastasis was identified by a DNA chip analysis by Ramaswamy et al.(**Table 3**) (Ramaswamy et al. 2003).

Table 3 The 17 -gene signature associated with metastasis

Gene	Gene name	Gene ID
Upregulated in metastases:		
SNRP1	Small nuclear ribonucleoprotein 1	A1032612
EIFAEL3	Elongation inhibition factor 4E-like 3	AF038957
HNRNPAB	Heterogeneous nuclear ribonucleoprotein A/B	M65028
DHPS	Deoxyhypusine synthase	U73262
PTTG	Securin	AA203476
COL1A1	Type 1 collagen, α1	Y15915
COL1A2	Type 1 collagen, α2	J03464
LAMB1	Lamin, β1	L37747
Downregulated in metastases:		
ACTG2	Actin, γ2	D00654
MYLK	Myosin light chain kinase	U49959
MYH11	Myosin heavy chain 11	AF001548
CLIN1	Calpon n1	D17408
HLA-DPB1	MHL class II, DPβ1	NM_002121
AUNX1	Auditory neuropathy, X-linked recessive 1	D43969
MT3	Metallothionein 3	S72043
NR4A1	Nuclear hormone receptor TR3	L13740
RBM5	RNA binding motif protein 5	AF391263

Several researchers have attempted the same approach and obtained gene sets that predict prognosis of patients with breast cancer. Sorile selected a set of 456 genes (Sorlie et al. 2003) (PNAS 2001). Van't Veer selected a set of 231 genes determining the prognosis and validated it using the independent data of 295 patients (van de Vijver et al. 2002). However, in relation to their selected genes, a few genes were overlapped. This discrepancy might be due to the following reasons: 1) validation in different clinical backgrounds, such as disease, histology, response criteria, and treatment, 2) difference in the assay methods used for RNA purification and in the methods used for gene amplification, and/or 3) difference in the analytical process used, such as standardizations and algorithms. How should future biomarker studies be considered? Future biomarker studies should include: 1) a prospective correlative study between markers and clinical features(survival and response e.g.), 2) barebones sample size, and 3) validation on another platform (Simon 2005).

Another problem is that the selected markers usually contain many functionally unknown markers. Therefore, it is difficult to discuss the implications of biomarkers without the availability of biological information. At the same time, it is necessary to analyze the functions of each biomarker, which requires much effort. Therefore, investigators should start biomarker (pharmacogenomic) studies in the early phase. Statistically, algorithms and data sets containing the biological information should be constructed. In addition, standardization of these analytical methods is essential. For clinical side application?, adequate prospective clinical studies are required. It is thus of utmost importance to establish better communication between clinical researchers, basic researchers and bio-statisticians from the planning stage.

References

Innocenti F, Undevia SD, Iyer L, Chen PX, Das S, Kocherginsky M, Karrison T, Janisch L, Ramirez J, Rudin CM, Vokes EE,Ratain MJ (2004) Genetic variants in the UDP-glucuronosyltransferase 1A1 gene predict the risk of severe neutropenia of irinotecan. J Clin Oncol 22: 1382-1388

McShane LM, Altman DG, Sauerbrei W, Taube SE, Gion M,Clark GM (2005) Reporting recommendations for tumor marker prognostic studies. J Clin Oncol 23: 9067-9072

Olaussen KA, Dunant A, Fouret P, Brambilla E, Andre F, Haddad V, Taranchon E, Filipits M, Pirker R, Popper HH, Stahel R, Sabatier L, Pignon JP, Tursz T, Le Chevalier T,Soria JC (2006) DNA repair by ERCC1 in non-small-cell lung cancer and cisplatin-based adjuvant chemotherapy. N Engl J Med 355: 983-991

Ramaswamy S, Ross KN, Lander ES,Golub TR (2003) A molecular signature of metastasis in primary solid tumors. Nat Genet 33: 49-54

Simon R (2005) Roadmap for developing and validating therapeutically relevant genomic classifiers. J Clin Oncol 23: 7332-7341

Sorlie T, Tibshirani R, Parker J, Hastie T, Marron JS, Nobel A, Deng S, Johnsen H, Pesich R, Geisler S, Demeter J, Perou CM, Lonning PE, Brown PO, Borresen-Dale AL,Botstein D (2003) Repeated observation of breast tumor subtypes in independent gene expression data sets. Proc Natl Acad Sci U S A 100: 8418-8423

van de Vijver MJ, He YD, van't Veer LJ, Dai H, Hart AA, Voskuil DW, Schreiber GJ, Peterse JL, Roberts C, Marton MJ, Parrish M, Atsma D, Witteveen A, Glas A, Delahaye L, van der Velde T, Bartelink H, Rodenhuis S, Rutgers ET, Friend SH,Bernards R (2002) A gene-expression signature as a predictor of survival in breast cancer. N Engl J Med 347: 1999-2009

Epidermal Growth Factor (EGF) Receptor Kinase Activity is Required for Tumor Necrosis Factor (TNF)-α Mediated Intestinal Epithelial Survival

Toshimitsu Yamaoka[1,3], D. Brent Polk[3], Tohru Ohmori[2], Sojiro Kusumoto[1], Tomohide Sugiyama[1], Takao Shirai[1], Masanao Nakashima[1], Kentaro Okuda[1], Takashi Hirose[1], Tsukasa Ohnishi[1], Naoya Horichi[1], and Mitsuru Adachi[1]

[1]1st Department of Internal Medicine, Showa University School of Medicine, 1-5-8 Hatanodai, Shinagawa-ku, Tokyo, 142-8666, Japan
[2]Institute of Molecular Biology, Showa University, 1-5-8 Hatanodai, Shinagawa-ku, Tokyo, 142-8666, Japan
[3]Department of Pediatrics, Cell and Developmental Biology, Vanderbilt University School of Medicine, 1025 MRBIV, Nashville, TN 37232-0696, USA

Summary. Epidermal growth factor receptor (EGFR) plays a key role in the regulation of normal cellular process and in the hyper-proliferative diseases such as cancer. Recently it has been demonstrated that TNF-induced apoptosis is augmented by inhibition of EGFR. In the present study we demonstrated that EGFR expression and its tyrosine kinase activity is required for TNF-induced cell survival using EGFR tyrosine kinas inhibitor, AG1478 in stable transfected colon intestinal epithelial $EGFR^{-/-}$ cells model transfected with either wild-type (wt), or kianse inactive (ki) EGFR. Furthermore, for the first time, we demonstrated that EGFR transactivation leads to other ErbB receptors through TNF induced transactivation of ErbB family receptors (EGFR, ErbB2 and ErbB4). Mutational analysis demonstrated that EGFR autophosphorylation sites of Y1068 and Y1086, a subsequent PI 3-kinase/Akt activation sites, is required for TNF mediated transactivation of EGFR, ErbB2 and 4 through Src kinase activity and cell survival. Like as Src kinase inhibitor, PI 3-kinase inhibitor enhanced TNF-induced apoptosis, but not MAP kinase inhibitor. From the present study we postulate that EGFR and EGFR tyrosine kinase involving Y1068 and Y1086 is critical in inducing TNF-initiated colon epithelial cell apoptosis through activation of the PI 3-kinase/AKT pathway. This novel observation has significant implications for understanding the role of EGFR regulation in maintaining intestinal epithelial cell homeostasis in an environment of chronic inflammation, injury and repair, such as inflammatory bowel disease and tumorigenesis.

Key words. EGFR, TNF, transactivation, apoptosis, Src

Introduction

Epidermal growth factor receptor (EGFR) tyrosine kinase activity is upregulated in many cancers and inflammations, through cytokines and growth factor stimulation. Tumor necrosis factor (TNF)-α is a potential cytokine, produced by many cell types in response to inflammation, injury, infection and environmental stress. TNF as well as causing necrotic cell death, also causes apoptotic cell death, cellular proliferation, differentiation, inflammation, tumorigenesis, and viral replication.

It has become clear that tumorigenesis is not only dependent on increase of growth, but also on deregulated apoptosis. Apoptosis can be initiated by variety of stimulation including growth factor deprivation, chemotherapeutic drugs, irradiation, activation of cell surface receptors of TNFR family and many more (Vaux & Strasser, 1996). Therefore increased resistance against apoptosis can be great importance for tumor development. Recent studies have linked ErbB signaling to protection from TNF-induced apoptosis (Fong, Sherwood, Mendelsohn, Lee, & Kozlowski, 1992; Zhou et al., 2000). However, the mechanism of tyrosine phosphorylation of the EGFR by TNF is unknown. EGFR transactivation has been reported in response to G-protein coupled receptor (GPCR) activation. Originally, Daub et al reported that the treatment of rat fibroblasts with lysophosphatidic acid (LPA), endothelin-1 (ET-1) or thrombin leads to rapid transient EGFR phosphorylation and subsequent activation of downstream signaling such as MAPK (Daub, Weiss, Wallasch, & Ullrich, 1996). Various studies demonstrated that GPCR-induced EGFR signal transactivation occurs in a variety of cell types, including vascular smooth muscle cells, human keratinocytes, primary mouse astrocytes and PC-12 cells (Daub, Wallasch, Lankenau, Herrlich, & Ullrich, 1997; Eguchi et al., 1998; Zwick et al., 1997). Possibly, there are two kind of GPCR-EGFR cross talk mechanism. One is a ligand-dependent and another is a ligand-independent intracellular signaling mechanism.

Prenzel et al were the first to demonstrate the ligand-dependent mechanism in EGFR signal transactivation by metalloprotease-mediated activation of the EGF-like ligand HB-EGF (Prenzel et al., 1999). Interestingly, Argust et al demonstrated that TNF transactivate EGFR via cleavage of TGF-α by TNF converting enzyme (TACE, ADAM17) and stimulates cell proliferation (Argast, Campbell, Brooling, & Fausto, 2004). Src-family tyrosine kinases have been suggested as both upstream and downstream mediators of the EGFR in GPCR-induced transactivation. Several studies have suggested that the presence of Src-family kinases upstream of the receptor in vascular smooth muscule cells, in immortalized hypothalamic neurons, and in LPA-stimulated COS-7 cells (Bokemeyer, Schmitz, & Kramer, 2000; Luttrell, Della Rocca, van Biesen, Luttrell, & Lefkowitz, 1997; Shah, Farshori, Jambusaria, & Catt, 2003). In contrast, other studies showed EGFR phosphorylation to be independent of Src activity in COS 7 and HEK-293 cells (Adomeit et al., 1999; Daub et al., 1997; Slack, 2000). Beside Src kinases, the protein kinase C (PKC) and the intracellular Ca^{+} concentration have been reported to be related with EGFR signal transactivation (Keely, Calandrella, & Barrett, 2000; Kodama et al., 2002; Soltoff, 1998). Using the antibodies against EGFR specific autophosphorylation sites, Y1068 and Y1173 EGFR were detected as EGFR transactivation site by angiotensin II, and Src-dependent phosphorylation of the EGFR Y845 is required for EGFR transactivation (Ushio-Fukai et al., 2001; Wu, Graves, Gill, Parsons, & Samet, 2002).

In this study, we determined the requirment of EGFR tyrosine kinase, especially Y1068 and Y1086 for EGFR transactivation by TNF. This EGFR transactivation mechanism is ligand-independent Src dependent intracellular signal, and this EGFR transactivation might regulate TNF-induced apoptosis via PI 3-kinase/Akt signaling.

Materials and Methods

Cell culture

EGFR-null mouse colon epithelial ($EGFR^{-/-}$MCE) cells were isolated from the colonic epithelium of EGFR-null heterozygous mice crossed with the Immorto mouse (Threadgill et al., 1995; Whitehead, VanEeden, Noble, Ataliotis, & Jat, 1993). Cells were maintained on collagen (Mediatech, Herndon, VA) coated plates (5 μg/cm^2) in RPMI 1640 supplemented with 5% FBS, 5U/ml IFN-γ (Intergen, Norcross, GA), 100U/ml penicillin and streptomycin, 5 μg/ml insulin, 5 μg/ml transferring, 5 ng/ml selenous acid (BD Biosciences, San Jose, CA) at 33°C (permissive condition). Prior to all experiments cells were serum starved for 16-18 hours in RPMI 1640 containing 0.5% FBS and 100U/ml penicillin and streptomycin at 37°C (nonpermissive conditions).

Antibodies, growth factors, and inhibitors

Rabbit monoclonal pY1068, and pY1173 EGFR, rabbit polyclonal phospho-AKT (Ser 473) and AKT antibodies, and horseradish peroxidase (HRP)-conjugated anti-rabbit secondary antibody were from Cell Signaling Technology (Beverly, MA). Anti-active ERK1/2 polyclonal antibody was from Promega Corp. (Madison, WI). Rabbit polyclonal EGFR and phospho-HER2 (tyr1248) antibodies were from Upstate (Billerica, MA). Rabbit polyclonal phospho-ErbB4 (tyr1162) was from Orbigen (San Diego, CA). Recombinant human EGF was a gift from Stanley Cohn (Vanderbilt University, Nashville, TN). EGFR tyrosin kinase inhibitor AG1478, MEK1/2 inhibitor U0126, phosphoinositide 3-kinase inhibitors wortmannin and LY294002, matrix metalloproteiases (MMPs) inhibitor GM6001, and TACE (TNF-α converting enzyme/ADAM17) inhibitor TAPI-1 were from Calbiochem (San Diego, CA).

Generation of stable cells

EGFR$^{-/-}$MCE cells were transfected with pcDNA3.1/Zeo vector contro (Invitrogen, Carlsbad, CA) according to the manufacture's protocol. Zeocin-seleced pools of cells were stained with anti-EGFR 528-PE antibody (50 μ per 5×10^{6} cells). PE (EGFR) positive cells were sorted at the V.A. Medica Center Flow Cytometry Resource Center (Nashville, TN) using a Becton-Dickinson FACSAria. Stable pool cells expressing EGFR were maintained i 200 μg/ml Zeocin.

siRNA and transient transfection

Mouse Src, Fyn and Yes SMARTpool siRNA were from Dharmacon RNA Technologies (Lafayette, CO). YAMC cells were transfected at70% confluence with either 50nM non-targeting siRNA or 50nM Src, Fyn and Ye SMARTpool siRNA using Lipofectamine 2000 (Invitrogen, Carlsbad, CA according to the manufacture's recommendations for siRNA transfections. 2 hour after transfection, the cells were starved for 16 -18 hours and western blo analysis was performed as described below.

Apoptosis assay

EGFR$^{-/-}$MCE cells were cultured on fibronectin-coated chamber glass slides and prepared as described above. Following treatment, apoptotic cells were labeled by Apoptag in situ apoptosis detection kits (Intergen, Purchase, NY) using terminal deoxynucleotidyltransferase for detection of positive cells according to manufacture's guidelines. Apoptotic cells were labeled by fluoresceinisothiocyanate-conjugated anti-digixigenin and 3, 3'-diaminobenzidine as substrate and then dehydrated and mounted using Vectashield mounting medium with DAPI.

Caspase activity was detected in living cells using the sulforhodamine multi-caspase activity kit (Biomol, Plymouth Meeting, PA) according to manufacture's guidelines. Active caspase enzyme in living cells was labeled with cell-permeable sulforhodamine-conjugated valylalanylaspartic acid fluoromethyl ketone, an inhibitor of caspase activity that binds to active caspase enzyme. Cells with increased caspase activity were detected by fluorescence microscopy. Positive cells were determined by counting at least 200 cells in randomly chosen fields and expressing them as percentage of the total number of cells counted.

Apoptosis was also detected by Annexin V-FITC Apoptosis Detection kit I (BD Pharmingen, San Jose, CA). Cells were harvested with Accutase (Innovative Cell Technologies, San Diego, CA), sequentially incubated with FITC-conjugated annexin-V and propidium iodide following the manufacturer's recommendations. Stained cells were analyzed with FACScan and Cell Quest software.

Cell lysates, SDS-PAGE, and Western blot analysis

Serum-starved cells were treated with murine TNF-a (Pepro Tech, Inc., Rocky Hill, NJ) or murine EGF (gift from Stanley Cohen, Vanderbilt University, Nashville, TN) at 37°C. For inhibitor studies, cells were treated with inhibitor or an equal volume of vehicle for 45 minutes at 37°C prior to TNF or EGF stimulation. After treatment, cell monolayers were rinsed twice with cold-PBS and then scraped into cell lysis buffer (50mM Tris-HCl (pH 7.4), 120mM NaCl, 1% NP-40) with protease and phosphatase 1 and 2 inhibitor cocktails (Sigma, St. Louis, MO). The scraped suspensions were centrifuged (14,000×g for 10 min) at 4°C, and the protein concentration was determined with the DC protein assay (Bio-Rad Laboratories, Hercules, CA). 20 – 30 μg of total lysate for each sample were separated by SDS-polyacrylamide gel electrophoresis and transferred to nitrocellulose membranes. The membranes were blocked in

5% nonfat dry milk in Tris Buffer saline (50mM Tris, 138mM NaCl, 2.7mM KCl, pH 8.0) with 0.05% Tween 20 (TBST) for 1hour. The membranes were incubated in primary antibody at 4°C overnight, washed in TBST (3 × 5 min), incubated in HRP conjugated secondary antibody for 1 hour, and then washed in TBST (5 × 10min). Bound to HRP was detected using Western Lightning enhanced chemiluminescence kit (Perkin-Elmer, Boston, MA).

Statistical analysis

Statistical significance in each study was determined using a paired two-sample student's t-test with a confidence level of 0.05. All data presented are representative of at least three repeat experiments.

Results

EGFR tyrosine kinase activity modulates TNF-induced apoptosis in intestinal epithelial cells

Role of EGFR in TNF mediated apoptosis was investigated using $EGFR^{-/-}$ MCE cells. We transfected with the cells with either human EGFR wild-type (wt), kinase-inactive (ki) K721R mutant, which blocks ATP binding to EGFR or vector only, and stable clonal lines were established (Coker, Staros, & Guyer, 1994). As shown in figure 1B, EGFR was found to be equally expressed in wt and ki EGFR, but not in vector cells. In multicaspase activity detection assay, TNF induced apoptosis was found to be 10 fold higher in cells transfected with ki EGFR or vector cells, compared to control transfected with wt EGFR (Fig. 1A). Similar results were obtained in TUNEL assay (Data was not shown). However, under the presence of PI 3-kinase/AKT inhibitor, TNF significantly enhanced apoptosis in wt EGFR cells. These results indicated that EGFR expression and EGFR tyrosine kinase activity down regulates the induction of apoptosis by TNF, through PI 3-kinase/AKT signal cascade.

EGFR expression and tyrosine kinase activity are required ErbB2 & 4 transactivation by TNF

EGFR expression and its tyrosine kinase activation escapes TNF-induced apoptosis, we thought that there should be some interaction between TNF and EGFR. We looked into EGFR transactivation by TNF in $EGFR^{-/-}$ MCE cells

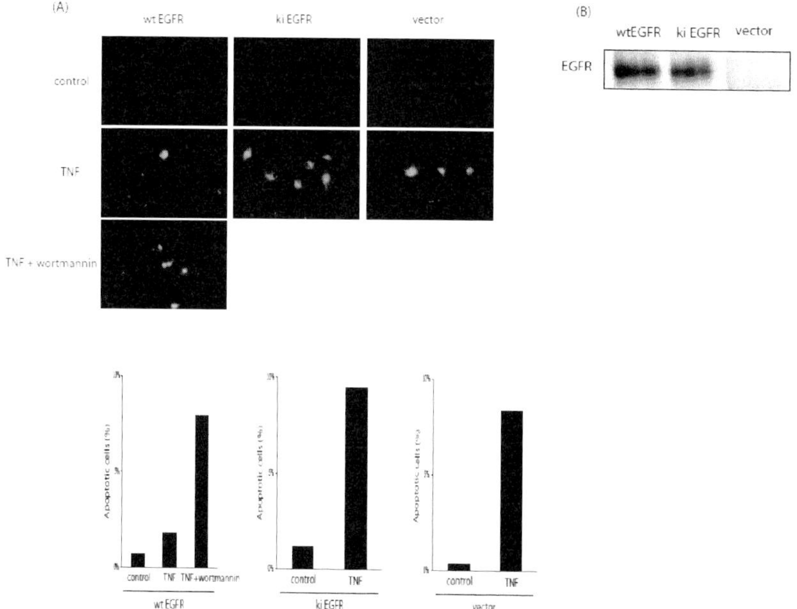

Fig.1 EGFR tyrosine kinase activity modulates TNF-induced apoptosis in intestinal epithelial cells. Cells were treated with TNF (100 ng/ml) for 6 hr in the presence or absence of 1-hr pretreatment with wortmannin. Caspase activity in living cells was detected using a multi-caspase activity assay kit and counterstained with DAPI. Images were obtained using fluorescence microscopy. The percentage of cells undergoing apoptosis is shown in the bar graph (A. Refer to color plates) Cellular lysates were prepared for Western blot analysis with anti-EGFR to determine EGFR expression levels (B).

by transfecting with wt or ki EGFR. TNF was found to transactivate EGFR in wt EGFR (Fig.2A). In addition, TNF transactivates ErbB2 and ErbB4 in wt EGFR cells, but we could not look into ErbB3 transactivation, as this $EGFR^{-/-}$ MCE cell does not express ErbB3. However, in ki EGFR cells, EGFR phosphorylation was not observed and also ErbB2 and ErbB4 receptors were not activated by TNF, indicating that TNF stimulates EGFR and its phosphorylation leads to activation of ErbB2 and ErbB4. This was further demonstrated by using EGFR tyrosine kinase inhibitor, AG1478, which blocked TNF mediated EGFR and ErbB2 and ErbB4 receptors phosphorylation (Fig.2B).

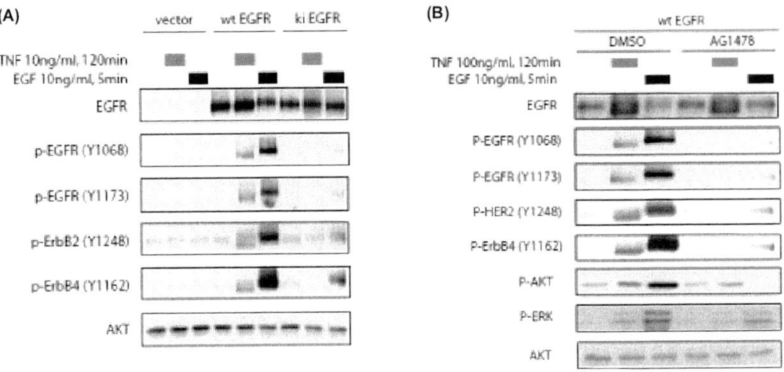

Fig. 2 EGFR expression and tyrosine kinase activity are required for ErbB2 and ErbB4 transactivation by TNF. Cells were treated with TNF (100 ng/ml) and EGF (10 ng/ml) for the indicated times in vector, wt EGFR and ki EGFR cells (A). wt EGFR cells were treated with TNF and EGF in the presence and absence of 1 hr pretreatment of AG1478 (150 nM) (B). Cellular lysates were prepared for Western blot analysis with anti-EGFR (EGFR), anti-phospho-EGFR tyr 1068 and tyr1173 (p-EGFR (Y1068) and p-EGFR (Y1173)), anti-phospho-ErbB2 tyr 1248 (p-ErbB2 (Y1248)), anti-phospho-ErbB4 tyr 1162 (p-ErbB4 (Y1162)), anti-AKT (AKT), anti-phospho-AKT Ser473 (P-AKT) and anti-phospho-ERK (P-ERK) antibodies.

TNF-induced transactivation of EGFR, Erbb2 and 4 required for EGFR Y1068 and Y1086

As indicated from above results, EGFR phosphorylation is required for TNF induced transactivation, we investigated the role of EGFR autophosphorylation site(s). Transactivation of EGFR by GPCRs is well established, but there is few information regarding the site-specific tyrosine phosphorylation and molecular regulation of the recepor. With the site directed mutagenesis, we mutated tyrosine (Y) to phenylalanine (F) in EGFR autophosphorylation site(s) and transfected our established stable clones of $EGFR^{-/-}$MCE.

TNF induced apoptosis was found to be significantly higher in Y1068F and Y1086F than other mutants. Furthermore, double mutant Y1068/1086F demonstrated higher induction of apoptosis than respective single mutants (Fig

3B). This indicated that Y1068 and Y1086 is important in TNF induced EGFR mediated apoptosis.

To investigate the role of these and other tyrosine residues and their phosphorylation, we investigated activation of receptors. TNF did not transactivate the phosphorylation of EGFR, ErbB2 and ErbB4 in cells transfected with Y1068F, Y1086F and Y1068/1086F. In Y992/1173F cells, TNF clearly activated EGFR and ErbB2, but not ErbB4 (Fig.3A). TNF was found to transactivate the phosphorylation of ErbB receptors (EGFR, ErbB2 and Erb4) in cells transfected with Y974F or Y1045F, and then did not enhanced apoptosis (Fig.3A and B). It is reported that the phosphorylation of EGFR tyrosine residues of 1068 and 1086 creates a binding site for Gab1 that recruits the p85 subunit of PI 3-kinase to the EGFR and leads to activation of AKT. These findings clearly demonstrated that Y1068 and Y1086 are required for TNF mediated transactivation of EGFR, ErbB2 and Erb4 and cell proliferation mediated through PI-3 kinase/AKT cascade.

ADAM/MMP inhibition does not block EGFR transactivation by TNF

TNF is known to activate matrixmetalloproteases (MMPs) that cleave membrane-integrated ErbB ligands (such as HB-EGF) and freeing them to bind to ErbBs. To clarify the mechanism of TNF induced activation of ErbB receptors and modulated apoptosis, we investigated the role of ADAM/MMPs. Inhibition analysis using GM6001, an inhibitor of ADAM/MMPs, in wt EGFR transfected cells exhibited transactivation of EGFR (Fig. 4). Similar results were obtained in ADAM17 specific inhibitor, TAPI-1 treatment (data was not shown) indicating that TNF mediated EGFR induced apoptosis does not require MMPs for ErbB ligand cleavage but through other signaling cascade.

TNF transactivates EGFR via Src and inhibition of Src enhanced TNF-induced apoptosis

EGFR is known to be transactivated through GPCRs (G-protein coupled receptors), which is known to indirectly activate Src and phosphorylates the intracellular domains of ErbBs on tyrosine residues. In order to clarify the signaling cascade that modulates TNF mediated EGFR induced apoptosis, we investigated Src cascade. We observed that treating with Src inhibitors, CGP77675, PP1 and PP2, enhanced TNF-induced apoptosis dose-dependently in TUNEL assay (Fig. 5C) indicating that Src cascade is responsible for TNF mediated EGFR induced apoptosis. It was found that these inhibitors inhibited

TNF mediated EGFR transactivation and AKT phosphorylation (Fig. 5A). To clarify that Src is responsible, we transfected with siRNA Src mixture, mixed

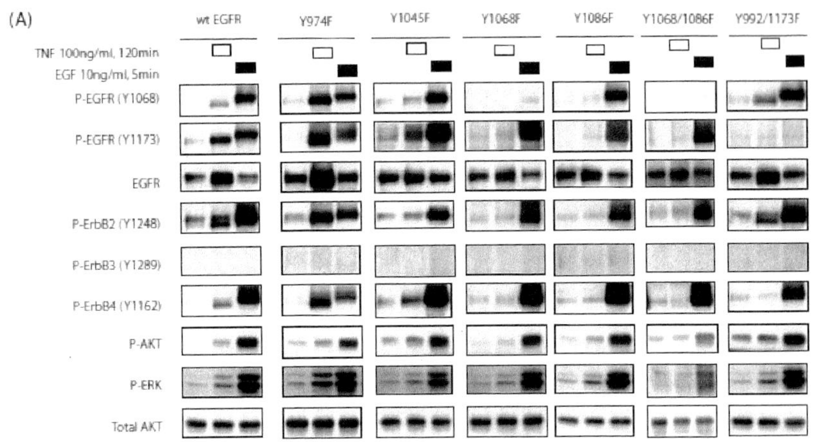

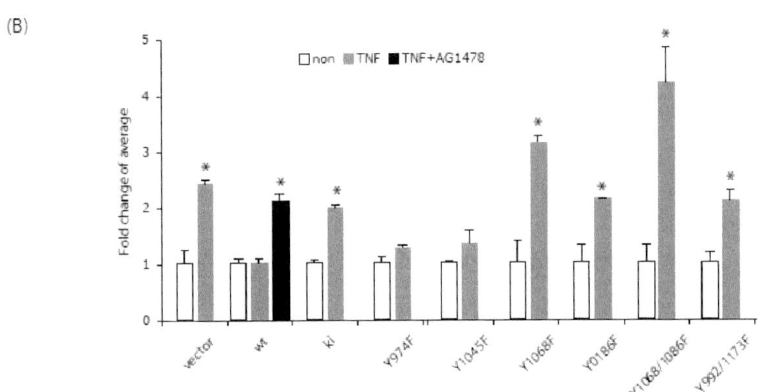

Fig. 3 TNF-induced transactivation of EGFR, ErbB2 and ErbB4 required for EGFR Y1068 and Y1086. Cellular lysates were prepared for Western blot analysis with anti-EGFR (EGFR), anti-phospho-EGFR tyr 1068 and tyr1173 (p-EGFR (Y1068) and p-EGFR (Y1173)), anti-phospho-ErbB2 tyr 1248 (p-ErbB2 (Y1248)), anti-phospho-ErbB4 tyr 1162 (p-ErbB4 (Y1162)), anti-AKT (AKT), and anti-phospho-AKT Ser473 (P-AKT) antibodies (A). The cells were treated with TNF (100ng/ml) for 6 hr and stained with Annexin V-FITC and propidium Iodide (Erlich et al.), and then analyzed

with flow cytometry within 1 hr. The percentage of apoptotic subpopulation Inhibition analysis using GM6001, an inhibitor of ADAM/MMPs, in wt EGFR with Fyn, Src and Yes, and found partially inhibition of EGFR transactivation that lead to ErbB2 and ErbB4 inhibition (Fig. 5B).

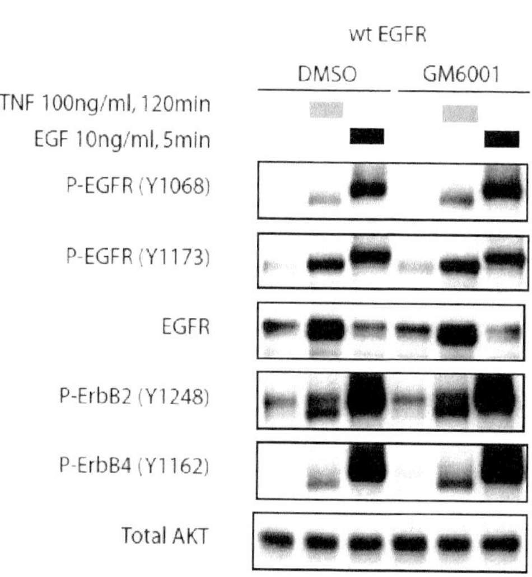

Fig. 4 ADAM/MMPs inhibition does not block EGFR transactivation by TNF. Cells were treated with TNF (100ng/ml) and EGF (10ng/ml) for indicated times in the presence and absence of 1 hr-pretreatment of ADAM/MMPs inhibitor, GM6001 (50μM). Cellular lysates were prepared for Western blot analysis with anti-EGFR (EGFR), anti-phospho-EGFR tyr 1068 and tyr1173 (p-EGFR (Y1068) and p-EGFR (Y1173)), anti-phospho-ErbB2 tyr 1248 (p-ErbB2 (Y1248)), anti-phospho-ErbB4 tyr 1162 (p-ErbB4 (Y1162)), and anti-AKT (AKT) antibodies.

Discussion

In the present study, we demonstrated the requirement of EGFR tyrosine kinase activity and the specificity of EGFR autophosphorylation site(s) regulating TNF-induced cell survivals. Although the mechanism(s) of EGFR transactivation by TNF is a poorly understood, we provide the evidence that TNF mediated EGFR induced cell survival is through Src cascade involving Y1068 and Y1086 EGFR tyrosine kinase activity.

EGFR expression and tyrosine kinase are upregulated in colon cancer and colitis than healthy colons (Malecka-Panas et al., 1997), that suggesting that the activation of EGFR is a response to these diseases in humans. In the

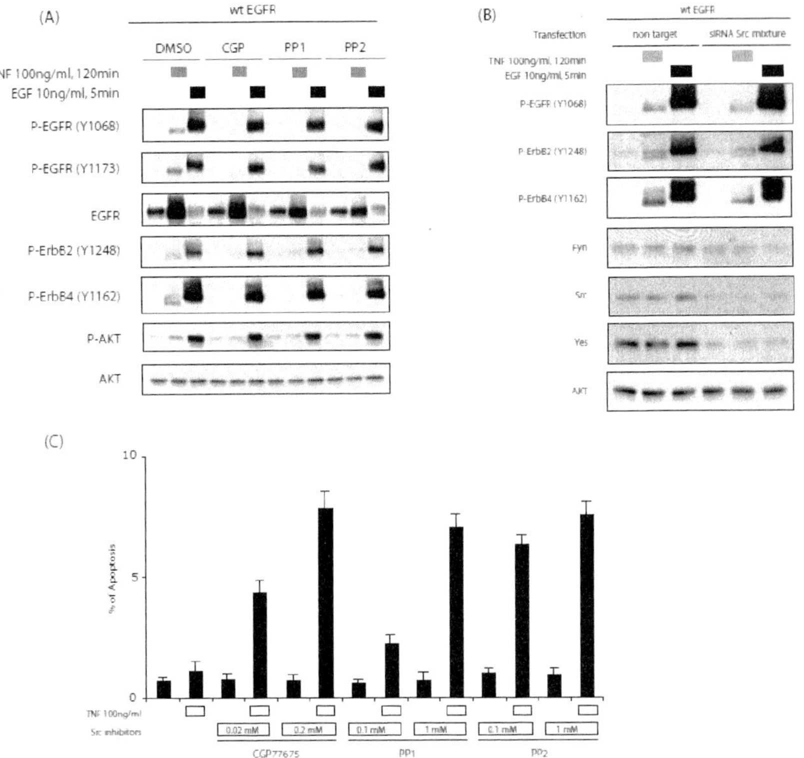

Fig. 5 TNF transactivates EGFR via Src and inhibition of Src enhanced TNF-induced apoptosis. wt EGFR cells were treated with TNF (100ng/ml) for 2 hr in the presence and absence of 1 hr pretreatment of Src inhibitors, CGP77675 (0.2μM), PP1 (1μM) and PP2 (1μM). Cellular lysates were prepared for Western blot analysis with anti-EGFR (EGFR), anti-phospho-EGFR tyr 1068 and tyr1173 (p-EGFR (Y1068) and p-EGFR (Y1173)), anti-phospho-ErbB2 tyr 1248 (p-ErbB2 (Y1248)), anti-phospho-ErbB4 tyr 1162 (p-ErbB4 (Y1162)), anti-AKT (AKT) and anti-phospho-AKT ser473 (P-AKT) antibodies. (A) wt EGFR cells were transfected with non-targeting siRNA as a control and siRNA Src mixture (Fyn, Src and Yes). Cells were treated with TNF (100 ng/ml) and EGF (10 ng/ml) as indicated times. Cellular lysates were prepared for Western blot analysis with anti-phospho-EGFR tyr 1068 and tyr1173 (p-EGFR (Y1068) and p-EGFR (Y1173)), anti-phospho-ErbB2 tyr 1248 (p-ErbB2 (Y1248)),

anti-phospho-ErbB4 tyr 1162 (p-ErbB4 (Y1162)), anti-Fyn (Fyn), anti-Src (Src), anti-Yes (Yes) and anti-AKT (AKT) antibodies. (B) Cells were treated with TNF (100ng/ml) for 6 hr in the presence and absence of 1 hr pretreatment of Src inhibitors, CGP77675, PP1 and PP2 as indicated concentrations, and fixed with TUNEL and DAPI staining. Apoptotic nuclei were labeled with FITC. Images were obtained using fluorescence microscopy. The percentage of cells undergoing apoptosis is shown in the bar graph (C).

present study, we hypothesized that EGFR tyrosine kinase regulates TNF-mediated colon epithelial cell survival. To clarify the mechanism of TNF mediated EGFR induced colon epithelial cell survival, we explored EGFR tyrosine kinase. Using EGFR$^{-/-}$ MCE cells model, and add back wild-type (wt), kinase-inactive (ki) EGFR, we found that EGFR expression and EGFR tyrosine kinase activity are required for TNF-mediated cell survival (Fig. 1A). However, Egeblad et al, demonstrated ErbB family receptors (EGFR, ErbB2, 3 and 4) do not require for regulation of TNF-induced cell death in MCF-7 breast cancer cells (Egeblad & Jaattela, 2000). But our data suggests a role for ErbB receptor activation in the induction of resistance to apoptosis. We found that ErbB receptor activation downregulates induction of apoptosis by TNF and conversely, loss of EGFR expression and EGFR tyrosine kinase inactivation enhanced TNF-induced apoptosis. Our results are in concordance to other studies have shown a correlation between the over-expression of ErbB receptors and resistance to TNF-induced cell death (Hoffmann, Schmidt, & Wels, 1998; Lichtenstein et al., 1990).

We found that Y1068/Y1086 residues are important in down regulating TNF induced apoptosis since Y1068F and Y1086F mutation enhanced apoptosis (Fig.3B). These results clearly demonstrates that Y1068 and Y1086 EGFR are required for TNF-mediated EGFR induced cell-survival, and each of Y1068 and Y1086 EGFR autophosphorylation sites are supportive for transducing the survival signals. Y1068 and Y1086 EGFR autophosphorylation sites create binding sites for Grb2, leading to activation of the MAPK/ERK cascade, and also create binding sites for Gab1, which recruits the p85 subunit of PI 3-kinase to the EGFR, leading to activation of Akt (Batzer, Rotin, Urena, Skolnik, & Schlessinger, 1994; Rodrigues, Falasca, Zhang, Ong, & Schlessinger, 2000; Rojas, Yao, & Lin, 1996). TNF has been reported to activate both an apoptotic cascade, as well as a cell survival signal through activation of PI 3-kinase/Akt pathway that suppresses TNF-induced apoptosis via NF-κB activation (Burow et al., 2000). Our site-specific inactive EGFR mutants, Y1068F and Y1086F are intrinsically suppressed PI 3-kinase/Akt activation, then TNF did not show clear stimulation of Akt, indicating that it subsequently did not activate NF-κB. As shown in Fig. 3B, the double inactive mutant EGFR, Y1068/1086F showed cumulative effect of single mutants in TNF-induced apoptosis. Furthermore, the blockade of PI 3-kinase/Akt signals with wortmannin in wt EGFR cells greatly enhanced TNF-mediated apoptosis. These results suggested that Y1068 and Y1086 of EGFR could be co-operatively stimulate PI 3-kinase/Akt signals and increase cell survival. These data support PI 3-kinase/AKT pathway should be mainly regulated TNF-induced apoptosis.

Erlich S et al showed that ErbB4 activation by neuregulin in PC-12 cells inhibits TNF-induced apoptosis and this effect of neuregulin is probably mediated by the PI 3-kinase/Akt activation (Erlich, Goldshmit, Lupowitz, & Pinkas-Kramarski, 2001). As results in Fig 3B, Y992/1173F EGFR cells are significantly increased TNF-induced apoptosis and ErbB4 (Y1162) was not transactivated by TNF and the lack of ErbB4 activation should be mediated TNF-induced apoptosis. These results indicate Y992/1173 EGFR is a transactivation site of ErbB4.

EGFR transactivation has been reported in response to G-protein couple receptor (GPCR) activation in several systems including intestinal epithelial cells (Keely, Uribe, & Barrett, 1998; Prenzel et al., 1999). Cross-talk between the EGFR and TNFRs has been suggested in a previous study (Izumi et al., 1994), however the mechanisms of tyrosine phosphorylation of the EGFR by TNF is unknown. There are two possible mechanisms of TNF mediated EGFR transactivation: ligand-dependent and independent mechanisms. The ligand-dependent mechanism, though poorly defined, is postulated to be ADAM (a integrated and matrixmetalloprotease)/MMPs (matrixmetalloproteases) activation by TNFR(s), and cleavage of membrane-tethered ErbB ligands (such as HB-EGF or TGF-α), thereby freeing them to bind to ErbB(s) (Argast et al., 2004; Chen et al., 2004). In the other ligand-independent mechanisms, GCPRs indirectly activate Src or other signal factors, which subsquently activates the intracellular domains of ErbB(s) on tyrosine residues (Luttrell et al., 1997). ADAM/MMPs inhibitor, GM6001 do not block EGFR transactivation (Fig. 4), however Src inhibitors, CGP77675, PP1 and PP2 clearly block the TNF mediated EGFR transactivation. Furthermore these Src inhibitors enhance TNF-induced apoptosis dose-dependently (Fig. 5C). These results clearly suggested ligand-independent mechanism and we have confirmed these effects by using siRNA Src, Fyn and Yes mixture. As shown in Fig. 5B, transfected with Src mixture (Src+Fyn+Yes) partially blocked TNF mediated EGFR transactivation by partially inhibiting each Src family members. Since EGFR transactivation is not found to be blocked when transfecting with individually individual siRNA for each Src family members, suggeating that Src family kinase members work in tandem to induce such effects.

In summary, we propose a novel mechanism of ErbB receptors interaction regulated through TNF-stimulated Src activation. These data provide an explanation for the requirement of EGFR activity for TNF-induced epithelial cell survival and suggesting potential further avenues of investigation into relationship between EGFR tyrosine residues and intracellular signaling. Furthermore, we provide the evidence that the tyrosine residues of 1068 and 1086 on EGFR serves as a molecular target regulating cell survival via PI 3-kinase/Akt pathway.

References

Adomeit, A., Graness, A., Gross, S., Seedorf, K., Wetzker, R., & Liebmann, C. 1999. Bradykinin B(2) receptor-mediated mitogen-activated protein kinase activation in COS-7 cells requires dual signaling via both protein kinase C pathway and epidermal growth factor receptor transactivation. Mol Cell Biol, 19(8): 5289-5297.

Argast, G. M., Campbell, J. S., Brooling, J. T., & Fausto, N. 2004. Epidermal growth factor receptor transactivation mediates tumor necrosis factor-induced hepatocyte replication. J Biol Chem, 279(33): 34530-34536.

Batzer, A. G., Rotin, D., Urena, J. M., Skolnik, E. Y., & Schlessinger, J. 1994. Hierarchy of binding sites for Grb2 and Shc on the epidermal growth factor receptor. Mol Cell Biol, 14(8): 5192-5201.

Bokemeyer, D., Schmitz, U., & Kramer, H. J. 2000. Angiotensin II-induced growth of vascular smooth muscle cells requires an Src-dependent activation of the epidermal growth factor receptor. Kidney Int, 58(2): 549-558.

Burow, M. E., Weldon, C. B., Melnik, L. I., Duong, B. N., Collins-Burow, B. M., Beckman, B. S., & McLachlan, J. A. 2000. PI3-K/AKT regulation of NF-kappaB signaling events in suppression of TNF-induced apoptosis. Biochem Biophys Res Commun, 271(2): 342-345.

Chen, W. N., Woodbury, R. L., Kathmann, L. E., Opresko, L. K., Zangar, R. C., Wiley, H. S., & Thrall, B. D. 2004. Induced autocrine signaling through the epidermal growth factor receptor contributes to the response of mammary epithelial cells to tumor necrosis factor alpha. J Biol Chem, 279(18): 18488-18496.

Coker, K. J., Staros, J. V., & Guyer, C. A. 1994. A kinase-negative epidermal growth factor receptor that retains the capacity to stimulate DNA synthesis. Proc Natl Acad Sci U S A, 91(15): 6967-6971.

Daub, H., Weiss, F. U., Wallasch, C., & Ullrich, A. 1996. Role of transactivation of the EGF receptor in signalling by G-protein-coupled receptors. Nature, 379(6565): 557-560.

Daub, H., Wallasch, C., Lankenau, A., Herrlich, A., & Ullrich, A. 1997. Signal characteristics of G protein-transactivated EGF receptor. Embo J, 16(23): 7032-7044.

Egeblad, M. & Jaattela, M. 2000. Cell death induced by TNF or serum starvation is independent of ErbB receptor signaling in MCF-7 breast carcinoma cells. Int J Cancer, 86(5): 617-625.

Eguchi, S., Numaguchi, K., Iwasaki, H., Matsumoto, T., Yamakawa, T., Utsunomiya, H., Motley, E. D., Kawakatsu, H., Owada, K. M., Hirata, Y., Marumo, F., & Inagami, T. 1998. Calcium-dependent epidermal growth factor receptor transactivation mediates the angiotensin II-induced mitogen-activated protein kinase activation in vascular smooth muscle cells. J Biol Chem, 273(15): 8890-8896.

Erlich, S., Goldshmit, Y., Lupowitz, Z., & Pinkas-Kramarski, R. 2001. ErbB-4 activation inhibits apoptosis in PC12 cells. Neuroscience, 107(2): 353-362.

Fong, C. J., Sherwood, E. R., Mendelsohn, J., Lee, C., & Kozlowski, J. M. 1992. Epidermal growth factor receptor monoclonal antibody inhibits constitutive receptor phosphorylation, reduces autonomous growth, and sensitizes androgen-independent prostatic carcinoma cells to tumor necrosis factor alpha. Cancer Res, 52(21): 5887-5892.

Hoffmann, M., Schmidt, M., & Wels, W. 1998. Activation of EGF receptor family members suppresses the cytotoxic effects of tumor necrosis factor-alpha. Cancer Immunol Immunother, 47(3): 167-175.

Izumi, H., Ono, M., Ushiro, S., Kohno, K., Kung, H. F., & Kuwano, M. 1994. Cross talk of tumor necrosis factor-alpha and epidermal growth factor in human microvascular endothelial cells. Exp Cell Res, 214(2): 654-662.

Keely, S. J., Uribe, J. M., & Barrett, K. E. 1998. Carbachol stimulates transactivation of epidermal growth factor receptor and mitogen-activated protein kinase in T84 cells. Implications for carbachol-stimulated chloride secretion. J Biol Chem, 273(42): 27111-27117.

Keely, S. J., Calandrella, S. O., & Barrett, K. E. 2000. Carbachol-stimulated transactivation of epidermal growth factor receptor and mitogen-activated protein kinase in T(84) cells is mediated by intracellular ca(2+), PYK-2, and p60(src). J Biol Chem, 275(17): 12619-12625.

Kodama, H., Fukuda, K., Takahashi, T., Sano, M., Kato, T., Tahara, S., Hakuno, D., Sato, T., Manabe, T., Konishi, F., & Ogawa, S. 2002. Role of EGF Receptor and Pyk2 in endothelin-1-induced ERK activation in rat cardiomyocytes. J Mol Cell Cardiol, 34(2): 139-150.

Lichtenstein, A., Berenson, J., Gera, J. F., Waldburger, K., Martinez-Maza, O., & Berek, J. S. 1990. Resistance of human ovarian cancer cells to tumor necrosis factor and lymphokine-activated killer cells: correlation with expression of HER2/neu oncogenes. Cancer Res, 50(22): 7364-7370.

Luttrell, L. M., Della Rocca, G. J., van Biesen, T., Luttrell, D. K., & Lefkowitz, R. J. 1997. Gbetagamma subunits mediate Src-dependent phosphorylation of the epidermal growth factor receptor. A scaffold for G protein-coupled receptor-mediated Ras activation. J Biol Chem, 272(7): 4637-4644.

Malecka-Panas, E., Kordek, R., Biernat, W., Tureaud, J., Liberski, P. P., & Majumdar, A. P. 1997. Differential activation of total and EGF receptor (EGF-R) tyrosine kinase (tyr-k) in the rectal mucosa in patients with adenomatous polyps, ulcerative colitis and colon cancer. Hepatogastroenterology, 44(14): 435-440.

Prenzel, N., Zwick, E., Daub, H., Leserer, M., Abraham, R., Wallasch, C., & Ullrich, A. 1999. EGF receptor transactivation by G-protein-coupled receptors requires metalloproteinase cleavage of proHB-EGF. Nature, 402(6764): 884-888.

Rodrigues, G. A., Falasca, M., Zhang, Z., Ong, S. H., & Schlessinger, J. 2000. A novel positive feedback loop mediated by the docking protein Gab1 and phosphatidylinositol 3-kinase in epidermal growth factor receptor signaling. Mol Cell Biol, 20(4): 1448-1459.

Rojas, M., Yao, S., & Lin, Y. Z. 1996. Controlling epidermal growth factor (EGF)-stimulated Ras activation in intact cells by a cell-permeable peptide mimicking phosphorylated EGF receptor. J Biol Chem, 271(44): 27456-27461.

Shah, B. H., Farshori, M. P., Jambusaria, A., & Catt, K. J. 2003. Roles of Src and epidermal growth factor receptor transactivation in transient and sustained ERK1/2 responses to gonadotropin-releasing hormone receptor activation. J Biol Chem, 278(21): 19118-19126.

Slack, B. E. 2000. The m3 muscarinic acetylcholine receptor is coupled to mitogen-activated protein kinase via protein kinase C and epidermal growth factor receptor kinase. Biochem J, 348 Pt 2: 381-387.

Soltoff, S. P. 1998. Related adhesion focal tyrosine kinase and the epidermal growth factor receptor mediate the stimulation of mitogen-activated protein kinase by the G-protein-coupled P2Y2 receptor. Phorbol ester or [Ca2+]i elevation can substitute for receptor activation. J Biol Chem, 273(36): 23110-23117.

Threadgill, D. W., Dlugosz, A. A., Hansen, L. A., Tennenbaum, T., Lichti, U., Yee, D., LaMantia, C., Mourton, T., Herrup, K., Harris, R. C., & et al. 1995.

Targeted disruption of mouse EGF receptor: effect of genetic background on mutant phenotype. Science, 269(5221): 230-234.

Ushio-Fukai, M., Griendling, K. K., Becker, P. L., Hilenski, L., Halleran, S., & Alexander, R. W. 2001. Epidermal growth factor receptor transactivation by angiotensin II requires reactive oxygen species in vascular smooth muscle cells. Arterioscler Thromb Vasc Biol, 21(4): 489-495.

Vaux, D. L. & Strasser, A. 1996. The molecular biology of apoptosis. Proc Natl Acad Sci U S A, 93(6): 2239-2244.

Whitehead, R. H., VanEeden, P. E., Noble, M. D., Ataliotis, P., & Jat, P. S. 1993. Establishment of conditionally immortalized epithelial cell lines from both colon and small intestine of adult H-2Kb-tsA58 transgenic mice. Proc Natl Acad Sci U S A, 90(2): 587-591.

Wu, W., Graves, L. M., Gill, G. N., Parsons, S. J., & Samet, J. M. 2002. Src-dependent phosphorylation of the epidermal growth factor receptor on tyrosine 845 is required for zinc-induced Ras activation. J Biol Chem, 277(27): 24252-24257.

Zhou, B. P., Hu, M. C., Miller, S. A., Yu, Z., Xia, W., Lin, S. Y., & Hung, M. C. 2000. HER-2/neu blocks tumor necrosis factor-induced apoptosis via the Akt/NF-kappaB pathway. J Biol Chem, 275(11): 8027-8031.

Zwick, E., Daub, H., Aoki, N., Yamaguchi-Aoki, Y., Tinhofer, I., Maly, K., & Ullrich, A. 1997. Critical role of calcium- dependent epidermal growth factor receptor transactivation in PC12 cell membrane depolarization and bradykinin signaling. J Biol Chem, 272(40): 24767-24770.

A Molecular Mechanism of Diminished Binding Activity between 15 bp Deletion Mutant EGFR and c-Cbl Ubiquitin Ligase

Tohru Ohmori[1,2], Takamichi Hosaka[1], Tomoko Kanome[2], Fumiko Inoue[2], Koichi Ando[1], Takashi Hirose[1], Tsuyoki Kadofuku[3], and Mitsuru Adachi[1]

[1]First Department of Internal Medicine, School of Medicine and [2]Institute of Molecular Oncology, Showa University, Hatanodai 1-5-8, Shinagawa-ku, Tokyo 142-8555, Japan

Summary. Gefitinib sensitivity in non-small-cell lung cancer (NSCLC) is associated with activating mutations in epidermal growth factor receptor (EGFR). It has been reported that autophosphorylation of the mutant EGFR is prolonged as compared with wild type EGFR (Lynch et al NEJM 2004). Previously, we reported that the mutant EGFR underwent less protein degradation due to diminished binding to c-Cbl ubiquitin ligase as compared to wild type EGFR. To clarify the mechanism, we examined EGFR-induced site specific phosphorylation of c-Cbl. c-Cbl residues Y700, Y731, and Y774 were all phosphorylated by wild type EGFR, but the mutant EGFR could phosphorylate Y774 but Y700 and Y731. We speculated the alteration of phosphorylation status of c-Cbl might reduce its binding to EGFR. To confirm this hypothesis, we constructed mutant c-Cbl plasmids and established stable transfectants using wild type and the mutant EGFR transfectants. As the results, 1) the mutations of Y700F and Y731F but Y774F reduced binding of wild type EGFR and c-Cbl. 2) all the three mutant c-Cbls showed less binding activity to the mutant EGFR. From these observations, it was suggested that the alteration of substrate specificity for c-Cbl and conformation change of the mutant EGFR might reduce its binding to c-Cbl.

Key words. EGFR, acquired resistance, gefitinib, non-small-cell lung cancer, c-Cbl

Introduction

A possible new paradigm for determining response to gefitinib has been reported, following the discovery that in-frame mutations of EGFR are well-correlated to the hyper-responsiveness to gefitinib in patients with

NSCLC (Lynch et al. 2004, Paez et al. 2004). These mutations were small, in-frame deletions or substitutions clustered around the ATP-binding site in exons 18, 19 and 21 of the EGFR (Lynch et al. 2004). We previously identified a 15-bp in-frame deletion in exon 19 of EGFR (2411 to 2425) in a gefitinib-hypersensitive NSCLC cell line, PC-9. Lynch *et al.* reported that ligand-induced autophosphorylation of mutant EGFR is significantly prolonged as compared with wild-type EGFR (Lynch et al. 2004), and the same observation was seen in PC-9 cells. Recent studies revealed that ligand binding to EGFR triggers the internalization and subsequent degradation of the activated receptor by lysosomal proteases and 26S proteasome (Yarden and Sliwkowski 2001). Ligand-induced polyubiquitinylation of EGFR is known as an essential factor in the down-regulation of activated EGFR. Ligand-activated EGFR is strongly coupled to the Cbl adaptor proteins Cbl-b and c-Cbl, E3 ubiquitin ligases. These proteins recruit ubiquitin-loaded E2 enzymes. In general, the polyubiquitinylation serves as a signaling for degradation of activated EGFR by 26S proteasome (Fig. 1) (de Melker et al. 2001, Levkowitz et al. 1998, Joazeiro et al. 1999, Waterman et al. 1999). These processes lead to a clearing of activated receptors from the cell surface, thereby attenuating the mitogenic signals (Wells et al. 1990). Previously, we reported that the mutant EGFR underwent less protein degradation due to diminished binding to c-Cbl ubiquitin ligase as compared to wild type EGFR and may cause the prolongation of its autophosphorylation (Ando et al. 2005).

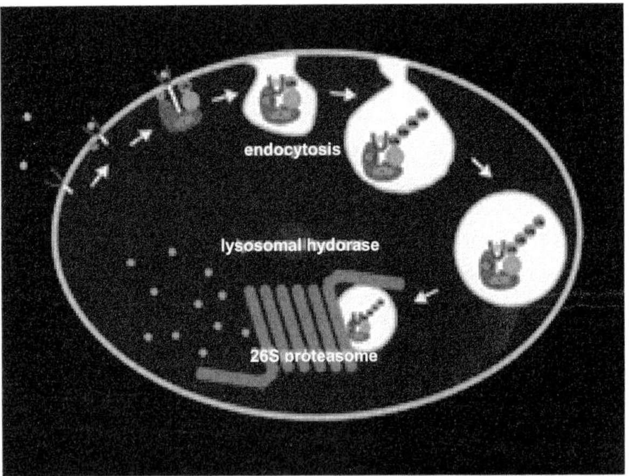

Fig. 1 Mechanism of EGFR-internalization & -degradation by c-Cbl Ligand-activated EGFR is strongly coupled to adaptor protein c-Cbl, E3 ubiquitin ligases. This protein recruits ubiquitin-loaded E2 enzymes. In general, the polyubiquitinylation serves as a signaling for degradation of activated EGFR by 26S proteasome.

c-Cbl is a substrate of protein tyrosine kinases (PTKs) and also undergoes tyrosine phosphorylation in response to EGFR activation (Take et al. 2000). It has been reported that the phosphorylation of c-Cbl Tyr-731 residue enhances its binding activity to EGFR and mediates following EGFR polyubiquitinylation and degradation (Soubeyran et al. 2002). To elucidate the mechanism of the diminished binding between c-Cbl and the mutant EGFR, we examined the alteration of c-Cbl-phosphorylation status in the mutant EGFR expressed cells and its influence to the binding activity.

Materials and Methods

Chemicals and antibodies

Gefitinib (Iressa) was donated by AstraZeneca (Wilmington, DE, USA). Anti-phospho-c-Cbl Tyr700 was purchased from BD Bioscience Pharmingen (San Jose, CA, USA), Tyr-731 and Tyr-774 antibodies and anti-phospho-EGFR (Tyr-845, Tyr-992, Tyr-1045, and Tyr-1068) antibodies were purchased from Cell Signaling Technology Inc. (Beverly, MA, USA). Anti-ubiquitin antibody and anti-β-actin antibody were purchased from Millipore (Billerica, MA, USA) and Ambion, Inc. (Austin, TX, USA), respectively. Other antibodies were purchased from Santa Cruz Biotech (SantaCruz, CA, USA). Chemicals were purchased from Sigma (St. Louis, MS, USA), unless otherwise mentioned.

Plasmid construction and cell lines

The construction of a mock vector (empty vector), wild-type EGFR expression vector, and the 15-bp deletion EGFR expression vector (delE746-A750 type deletion (Arao et al. 2004)), was described in detail in a previous paper (Koizumi et al. 2005). The stable transfectant HEK293 cell lines were kindly donated by Dr. Fukumoto (National Cancer Center, Shien Laboratory). The transfectant HEK293 cell lines, named 293_mock, 293_pEGFR and 293_pΔ15, contained the empty vector, wild-type EGFR and the 15-bp in-frame deletion mutant EGFR, respectively. Human c-Cbl expression vector was constructed using pcDNA3.1 and Y700F, Y731F, Y774F mutant c-Cbl expression vectors were constructed by QuickChange Stite-Directed Mutagenesis kit (Fig. 2). 293_mock, 293_pEGFR and 293_pΔ15 cells were transfected c-Cbl plasmids by Nucleofector kit (Amaxa biosystems) and stable transfectants were established.

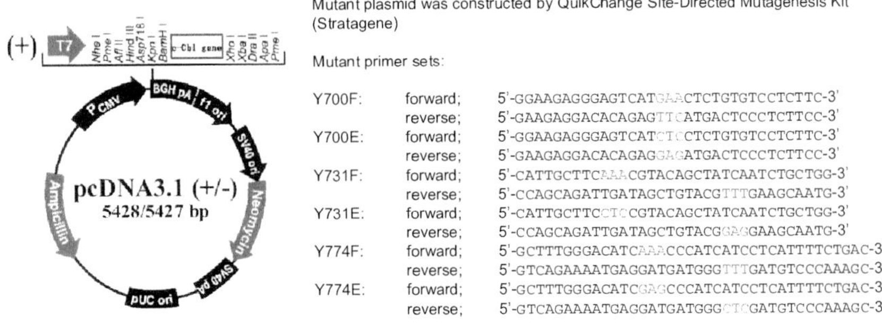

Transfection: Nucleofector™ Kit V (Amaxa biosystems)
Cell selection: 1 mg/ml Neomycin

Stable transfectants of 293_pEGFR and 293_pΔ15 cells were cloned and the established cell lines were named as follows:

293_pEGFR/Y700F, 293_pEGFR/Y731F 293_pEGFR/Y774F
293_pΔ15/Y700F, 293_pΔ15/Y731F, 293_pΔ15/Y774F

Fig. 2 Construction of mutant c-Cbl expression vectors

Immunoblot analysis

Cells were exposed to 100 ng/ml of TGF-α for indicated time periods and washed twice with ice-cold PBS. Immunoblotting and immunoprecipitation analysis was performed as described previously (Ando et al. 2005).

Ligand internalization assay and degradation assay

The EGFR internalization assay and the degradation assay were basically performed as described elsewhere, with minor modifications (Levkowitz et al. 1998, Sorkin et al. 1993, Kornilova et al. 1996).

Results and Discussion

Somatic mutations such as 15 bp deletion in exon 19 and point mutation in exon 21 are frequently found in gefitinib-hypersensitive patients. It has been reported these mutant EGFRs showed significantly prolonged autophosphorylation after ligand exposure as compared with wild type EGF receptor (Lynch et al. 2004). To elucidate the mechanism of this prolongation, we have examined the difference of EGFR-degradation activity between wild type EGFR and 15 bp deletion mutant EGFR(Akca et al. 2006). Previously, we established wild type EGFR and 15 bp

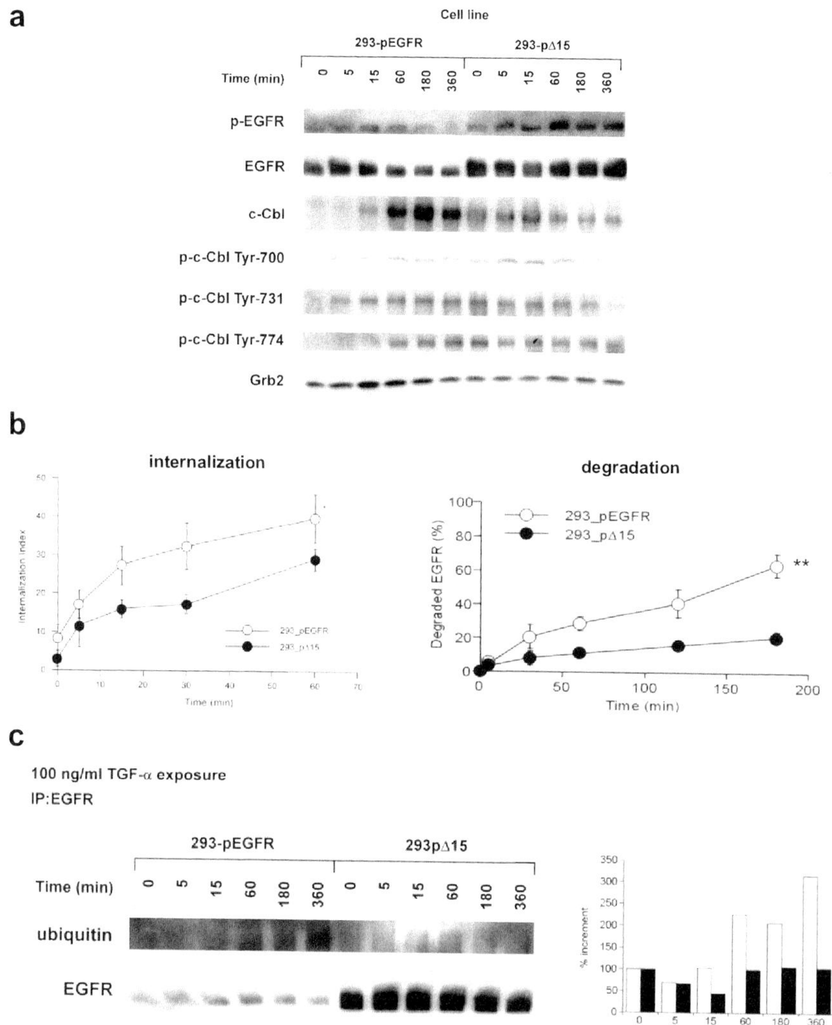

Fig. 3 c-Cbl binding and EGFR-degradation activities in the in-frame 15-bp deletion mutant EGFR transfectant cell line. (a) Cells were exposed to 100 ng/ml of TGF-α at 37℃. Samples were immunoprecipitated (IP) with anti EGFR antibody. (b) Cells were exposed to 2 ng/ml of ^{125}I-EGF for 1 h at 4℃and maintained at 37℃. means ± SD (bar) *P=0.094, **P<0.0001 (c) Sample was IP with anti EGFR antibody and IB by anti ubiquitin antibody. The graph indicates the degree of ubiquitinylation measured with a densitometer. Open column: 293_pEGFR. Closed column: 293_pΔ15

.deletion mutant EGFR stable transfectant cell lines usign HEK293 cells, and named 293_pEGFR and 293_pΔ15, respectively. The prolongation of mutant EGFR autophosphorylation after TGF-α exposure was confirmed in 293_pΔ15 cells, as compared with 293_pEGFR (Fig. 3a).

It has been reported that ligand binding to EGFR enhances endocytosis of the ligand-receptor complex (Yarden and Sliwkowski 2001) and that the degradation of this complex attenuates EGFR-mediated signal transduction (Wells et al. 1990, Jackson et al. 2000). Considering these observations, we speculated that the difference in the EGFR-turnover after the ligand exposure could contribute to the expansion of EGFR autophosphorylation. For the purpose of measuring EGFR-turnover, several kinetic assays were performed using ^{125}I-EGF. Fig. 3b shows EGFR internalizaion and degradation activities in these 2 cell lines, measured by ^{125}I-EGF. After 60 min incubation with 100 ng/ml TGF-α, the internalization activity in 293_ pΔ15 cells decreased to about 70% of that in 293_pEGFR cells. The degraded EGFR rates in 293_pEGFR cells and 293_pΔ15 cells were 63.9 ± 6.6% and 21.4 ± 3.0%, respectively. The degradation activity of the mutant EGFR was significantly decreased to about 30% of that of wild type EGFR. Moreover, significant ubiquitinylation of EGFR was observed in 293_pEGFR cells, but was rarely observed in 293_pΔ15 cells (Fig. 3c). In 293_pEGFR cells, EGFR-bound c-Cbl protein was significantly increased after 15 min exposure of TGF-α(Fig. 3a). Because this binding was correlated with the time course of phosphoryated EGFR downregulation, the degradation of wild type EGFR was thought to be triggered by c-Cbl binding in this cell line. In case of the mutant EGFR transfectant, the autophosphorylation of EGFR was observed after TGF-α exposure, and this activation was prolonged until 3 hours after the ligand exposure. In contrast with wild type EGFR, although clear autophosphorylation of mutant EGFR was detected, no enhancement of the c-Cbl binding was observed in 293_pΔ15 cells.

It has been reported that activated EGFR lead to c-Cbl binding directly through phosphorylation of Tyr-1045 residue in EGFR and indirectly through Grb2 binding. The mutations in EGFR that were thought to contribute to the gefitinib sensitivity, were located around ATP-binding site, and this site was distinct from the c-Cbl binding sites Tyr-1045 (Levkowitz et al. 1999), and Grb2 binding sites, Tyr-1068 and Tyr-1086 (Batzer et al. 1994, Jiang and Sorkin 2003). The phosphorylation status of the Tyr-1045 residue in EGFR is important for the c-Cbl binding to EGFR, and the replacement of the Tyr-1045 with a phenylalanine reduces ligand-induced downregulation and receptor ubiquitinylation (Levkowitz et al. 1999). We examined an immunoblotting using site specific phosphor-EGFR antibodies to determine if the phosphorylation status of EGFR was different between wild type EGFR and the mutant EGFR. In terms of the results, there was no difference in the phosphorylation status of Tyr-1045 in 293_pEGFR cells and 293_pΔ15 cells, moreover, phosphorylation of Tyr-1068 residue was more enhanced in 293_pΔ15 (Fig. 4b). Therefore, the conformation change in this mutant EGFR does not influence the phosphorylation status of these residues, and this phosphorylation does not contribute to the difference in c-Cbl binding between wild type EGFR and the mutant EGFR.

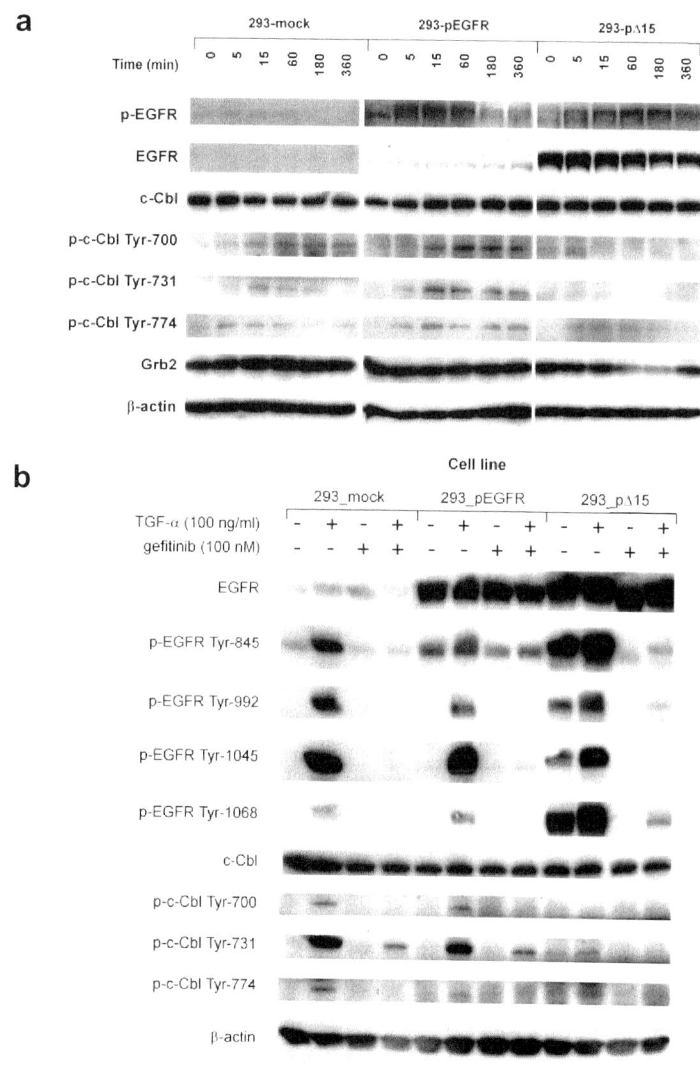

Fig. 4 Duration of EGFR autophosphorylation and EGFR expression in the in-frame 15-bp deletion mutant EGFR transfectant. (a)Cells were exposed to 100 ng/ml of TGF-α. After the treatment, cells were lysed and resolved by 10% SDS-PAGE. β-actin was detected as an internal control. (b)Cells were exposed to100 ng/ml TGF-α with/without 100 nM gefitinib. β-actin was detected as an internal control.

The other causative possibility for the alteration of c-Cbl binding is the phosphorylation status of c-Cbl by EGFR. c-Cbl is a substrate of EGFR tyrosine kinase and the E3 activity of c-Cbl is regulated by its tyrosine phosphorylation (Levkowitz et al. 1999). It was reported that the Tyr-731 residue in c-Cbl is a key site for the activation of its ubiquitin ligase activity (Levkowitz et al. 1999, Kassenbrock and Anderson 2004). Since the stable transfectants expressing the mutant EGFRs alter downstream substrate specificity (Sordella et al. 2004), we expected that a conformational change in the mutant EGFR possibly modulated the phosphorylation status of c-Cbl. To evaluate the effect of c-Cbl phosphorylation status on the diminished binding between mutant EGFR and c-Cbl, we examined site specific c-Cbl phosphorylation by respective phospho tyrosine specific antibodies. Y774 residue of c-Cbl was phosphorylated in both 293_pEGFR and 293_pΔ15 cell lines (Fig. 4a). In contrast, Y700 and Y731 residues of this protein were phosphorylated in 293_pEGFR cells, however, these residues were non or rarely phosphorylated in 293_pΔ15 cells. The phosphorylation of these residues were well correlated with the autophosphorylation of EGFR and were inhibited by 100 nM of gefitinib in 293_pEGFR, the phosphorylation of these residues was thought to be mediated by EGFR (Fig. 4b). From these observations, we speculated that mutant EGFR altered substrates specificity and the modification of c-Cbl phosphorylation status may decrease its binding activity to EGFR.

To confirm this hypothesis, we constructed mutant c-Cbl expression vectors, that were replaced the tyrosine residues of c-Cbl, Y700, Y731, and Y774, by phenylalanine (indicated F, dominant negative). 293_pEGFR and 293_pΔ15 cells were transfected with the mutant c-Cbl plasmids by an electroporation. The stable transfectants were cloned and designated, 293_pEGFR/Y700F, 293_pEGFR/Y731F, 293_pEGFR/Y774F, 293_pΔ15/Y700F, 293_pΔ15/Y731F, and 293_pΔ15/Y774F, respectively. The stable transfectant clones overexpressed mutant c-Cbl protein as compared with respective control cells and expression of c-Cbl levels were the mostly same in these transfectants (Fig. 5a). The expressed c-Cbl protein did not have any influence on the expression of EGFR protein at the non-stimulate condition. Using of these transfectants, we examined the binding activity between EGFR and mutant c-Cbl after TGF-α stimulation. In 293_pEGFR/c-Cbl cells, EGFR-bound wild type c-Cbl protein was significantly increased after 15 min exposure of TGF-α (Fig. 5b). In case of mutant c-Cbl protein, the replacement of Y700 and Y731 residues by phenylalanine decreased c-Cbl binding to wild type EGFR. Whereas the replacement of Y774 did not have any influence on the binding. From these observations, the phosphorylation of Y700 and Y731 residues by EGFR thought to have an important role for the binding between EGFR and c-Cbl. In 293_pΔ15/c-Cbl cells, this is somewhat discrepant data, but the overexpressed wild type c-Cbl was able to bind to the mutant EGFR, moderately. In contrast, all the mutant c-Cbl proteins included Y774F significantly decreased binding activity to EGFR. From these observations, it is thought that the reduced phosphorylation of c-Cbl Y700 and Y731 residues due to the alteration of substrate specificity as well as conformation change in the mutant EGFR might reduce the binding activity between the mutant EGFR and c-Cbl (Fig. 6).

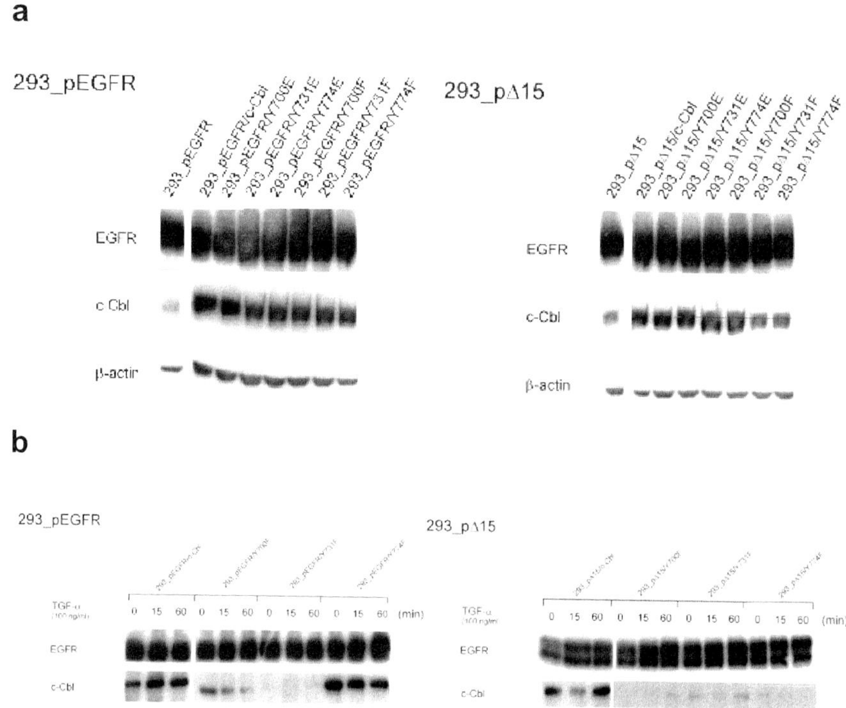

Fig. 5 Effect of phosphorylation status of c-Cbl on its binding activity to EGFR (a) 293_pEGFR cells and 293_pΔ15 cells were transfected c-Cbl plasmids. The expression of c-Cbl protein in the stable transfectants was demonstrated by an immunoblotting. (b) To examine the mutant c-Cbl binding to EGFR, the stable transfectants were exposed to 100 ng/ml of TGF-α for the indicated time periods at 37℃. Samples were IP with anti EGFR antibody.

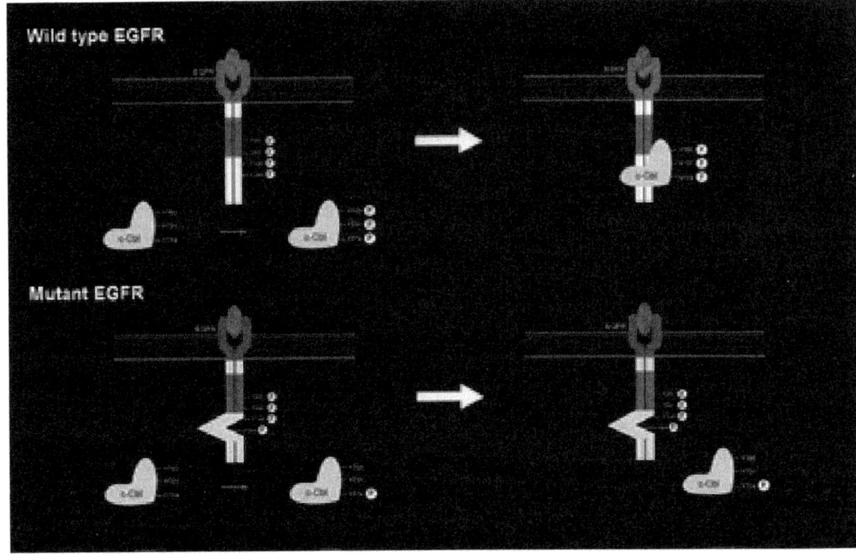

Fig. 6 Working hypothesis of the reduced binding between the mutant EGFR and c-Cbl From our observations, it is thought that both of the modulation of phosphorylation status in c-Cbl and the conformation change in the mutant EGFR may reduce the binding activity between the mutant EGFR and c-Cbl. Refer to color plates.

Acknowledgements

This work was supported in part by a Grant-in-Aid for the High-Technology Research Center Project from the Ministry of Education, Science, Sports and Culture of Japan, a Showa University Grant-in-Aid for Innovative Collaborative Research Projects and by a Special Research Grant-in-Aid for Development of Characteristic Education from the Japanese Ministry of Education, Culture, Sports, Science and Technology of Japan.

References

Akca H, Tani M, Hishida T, Matsumoto S,Yokota J (2006) Activation of the AKT and STAT3 pathways and prolonged survival by a mutant EGFR in human lung cancer cells. Lung Cancer 54: 25-33

Ando K, Ohmori T, Inoue F, Kadofuku T, Hosaka T, Ishida H, Shirai T, Okuda K, Hirose T, Horichi N, Nishio K, Saijo N, Adachi M,Kuroki T (2005) Enhancement of sensitivity to tumor necrosis factor alpha in non-small cell lung cancer cells with acquired resistance to gefitinib.

Clin Cancer Res 11: 8872-8879

Arao T, Fukumoto H, Takeda M, Tamura T, Saijo N,Nishio K (2004) Small in-frame deletion in the epidermal growth factor receptor as a target for ZD6474. Cancer Res 64: 9101-9104

Batzer AG, Rotin D, Urena JM, Skolnik EY,Schlessinger J (1994) Hierarchy of binding sites for Grb2 and Shc on the epidermal growth factor receptor. Mol Cell Biol 14: 5192-5201

Jackson PK, Eldridge AG, Freed E, Furstenthal L, Hsu JY, Kaiser BK,Reimann JD (2000) The lore of the RINGs: substrate recognition and catalysis by ubiquitin ligases. Trends Cell Biol 10: 429-439

Jiang X,Sorkin A (2003) Epidermal growth factor receptor internalization through clathrin-coated pits requires Cbl RING finger and proline-rich domains but not receptor polyubiquitylation. Traffic 4: 529-543

Joazeiro CA, Wing SS, Huang H, Leverson JD, Hunter T,Liu YC (1999) The tyrosine kinase negative regulator c-Cbl as a RING-type, E2-dependent ubiquitin-protein ligase. Science 286: 309-312

Kassenbrock CK,Anderson SM (2004) Regulation of ubiquitin protein ligase activity in c-Cbl by phosphorylation-induced conformational change and constitutive activation by tyrosine to glutamate point mutations. J Biol Chem 279: 28017-28027

Koizumi F, Shimoyama T, Taguchi F, Saijo N,Nishio K (2005) Establishment of a human non-small cell lung cancer cell line resistant to gefitinib. Int J Cancer 116: 36-44

Kornilova E, Sorkina T, Beguinot L,Sorkin A (1996) Lysosomal targeting of epidermal growth factor receptors via a kinase-dependent pathway is mediated by the receptor carboxyl-terminal residues 1022-1123. J Biol Chem 271: 30340-30346

Levkowitz G, Waterman H, Ettenberg SA, Katz M, Tsygankov AY, Alroy I, Lavi S, Iwai K, Reiss Y, Ciechanover A, Lipkowitz S,Yarden Y (1999) Ubiquitin ligase activity and tyrosine phosphorylation underlie suppression of growth factor signaling by c-Cbl/Sli-1. Mol Cell 4: 1029-1040

Levkowitz G, Waterman H, Zamir E, Kam Z, Oved S, Langdon WY, Beguinot L, Geiger B,Yarden Y (1998) c-Cbl/Sli-1 regulates endocytic sorting and ubiquitination of the epidermal growth factor receptor. Genes Dev 12: 3663-3674

Lynch TJ, Bell DW, Sordella R, Gurubhagavatula S, Okimoto RA, Brannigan BW, Harris PL, Haserlat SM, Supko JG, Haluska FG, Louis DN, Christiani DC, Settleman J,Haber DA (2004) Activating mutations in the epidermal growth factor receptor underlying responsiveness of non-small-cell lung cancer to gefitinib. N Engl J Med 350: 2129-2139

Paez JG, Janne PA, Lee JC, Tracy S, Greulich H, Gabriel S, Herman P, Kaye FJ, Lindeman N, Boggon TJ, Naoki K, Sasaki H, Fujii Y, Eck MJ, Sellers WR, Johnson BE,Meyerson M (2004) EGFR mutations in lung cancer: correlation with clinical response to gefitinib therapy. Science 304: 1497-1500

Sordella R, Bell DW, Haber DA,Settleman J (2004) Gefitinib-sensitizing EGFR mutations in lung cancer activate anti-apoptotic pathways. Science 305: 1163-1167

Sorkin A, Di Fiore PP,Carpenter G (1993) The carboxyl terminus of epidermal growth factor receptor/erbB-2 chimerae is internalization impaired. Oncogene 8: 3021-3028

Soubeyran P, Kowanetz K, Szymkiewicz I, Langdon WY,Dikic I (2002) Cbl-CIN85-endophilin complex mediates ligand-induced downregulation of EGF receptors. Nature 416: 183-187

Take H, Watanabe S, Takeda K, Yu ZX, Iwata N,Kajigaya S (2000) Cloning and characterization of a novel adaptor protein, CIN85, that interacts with c-Cbl. Biochem Biophys Res Commun 268: 321-328

Waterman H, Levkowitz G, Alroy I,Yarden Y (1999) The RING finger of c-Cbl mediates desensitization of the epidermal growth factor receptor. J Biol Chem 274: 22151-22154

Wells A, Welsh JB, Lazar CS, Wiley HS, Gill GN,Rosenfeld MG (1990) Ligand-induced transformation by a noninternalizing epidermal growth factor receptor. Science 247: 962-964

Yarden Y,Sliwkowski MX (2001) Untangling the ErbB signalling network. Nat Rev Mol Cell Biol 2: 127-137

de Melker AA, van der Horst G, Calafat J, Jansen H,Borst J (2001) c-Cbl ubiquitinates the EGF receptor at the plasma membrane and remains receptor associated throughout the endocytic route. J Cell Sci 114: 2167-2178

Part II

Inflammation, Angiogenesis, and Tumor Progression

TGFβ-Mediated Epithelial Mesenchymal Transition and Metastasis in Skin and Head-and-Neck Cancer

Sophia Bornstein[1], Gang-Wen Han[1,2], Kristina Hoot[1], Shi-Long Lu[1,2], Xiao-Jing Wang[1,2]

[1]Departments of Otolaryngology, Cell & Developmental Biology, and Dermatology, Oregon Health & Science University, Portland, Oregon.
[2]Departments of Pathology and Otolaryngology, University of Colorado Denver, Aurora, Colorado

Summary. In human skin and head and neck cancers, cancer cells frequently overexpress TGFβ but decrease expression of TGFβ signaling components. We generated several genetically engineered mouse models to investigate how these changes affect cancer progression. Overexpression of TGFβ1 in the skin or head and neck epithelia rapidly induced inflammation, suggesting that inflammation is a direct effect of TGFβ1 overexpression. Given the importance of inflammation in cancer development, our data suggest that in certain tissue types, TGFβ1-induced inflammation may override its tumor suppressive effect, even at early stages of carcinogenesis. This notion is further supported by our finding that deletion of TGFβ type II receptor (TGFβRII) in head and neck epithelia resulted in an elevated

---

*Correspondence to: Xiao-Jing Wang, M.D. & Ph.D., University of Colorado Denver Health Sciences Center, Bldg. RC1-N, Rm P28-5128, Mail Box 8104, 12800 E.19th Ave, Aurora, CO 80045 Tel: (303) 724-3001; FAX: (303) 724-3712; Email: XJ.Wang@UCHSC.edu

Abbreviations: TGFβ, transforming growth factor beta; SCC, squamous cell carcinoma; TGFβRII, TGFβ type II receptor; TGFβRI, TGFβ type I receptor; R-Smads, receptor-specific Smads; Co-Smad, common Smad; I-Smads, inhibitory Smads; ALK, activin receptor-like kinase; PAI-1, plasminogen activator inhibitor I; MMP, matrix metalloproteinases; TPA, phorbol 12-myristate 13-acetate.

endogenous TGFβ1, which was correlated with severe inflammation and angiogenesis in head and neck tissue. In combination with an activated Ras oncogene in TGFβRII null epithelial cells, these mice developed head and neck squamous cell carcinoma (HNSCC). Considering that most late-stage HNSCC cells are resistant to TGFβ1-mediated growth inhibition via loss of TGFβRII or other molecular alterations, our study suggests that inhibition of the effect of TGFβ1 on tumor stroma in combination with therapy targeting cancer epithelia may provide an effective therapeutic stragegy for HNSCC. Our study also revealed that dominant negative TGFβRII blocks TGFβ1-mediated EMT but not metastasis. Consistent with this finding, deletion of a common TGFβ signaling mediator, Smad4, in keratinocytes resulted in SCC formation but not EMT-type tumors. This result suggests that Smad4 loss abrogates TGFβ-mediated EMT, but not TGFβ-mediated invasion, the latter of which may be mediated by Smad3. This notion is further supported by our findings that Smad3 knockout mice are resistant to experimental skin carcinogenesis, and that Smad3 is rarely lost in SCCs. Taken together, our study suggests that SCCs that have both increased TGFβ1, and reduced TGFβRII or Smad4 in tumor epithelia will have a poorer prognosis than those with TGFβ1 overexpression alone or loss of TGFβRII/Smad4 alone.

Key words. transforming growth factor beta, skin carcinogenesis, epidermis, Smads, inflammation

Introduction

Transforming growth factor beta (TGFβ) represents a family of multifunctional cytokines that play a pivotal role in the maintenance of tissue homeostasis through regulation of biological processes including cell growth, differentiation, apoptosis, extracellular matrix formation, inflammation, and angiogenesis (Feng and Derynck, 2005). The TGFβ superfamily consists of three major subfamilies: TGFβ, activins/inhibins, and bone morphogenetic proteins (BMPs). These family members signal through type II and type I transmembrane serine/threonine kinase receptors. Three TGFβ isoforms,

TGFβ1, -2, and –3, have been identified in mammals with TGFβ1 being the most widely studied (Li et al., 2003; Wang, 2001). The Smad transcription factors were initially identified as TGFβ intracellular signaling mediators. Smads are divided into three groups: R-(receptor-activated) Smads, co-(common) Smad, and I-(inhibitory) Smads (Massague and Gomis, 2006). R-Smads include Smad1, -5, and -8, which are for BMP signaling, and Smad2 and -3, which are for TGFβ/activin signaling. When a TGFβ superfamily ligand binds its specific type II and type I receptor complex, the kinase domain of the type I receptor binds and phosphorylates R-Smads. Phosphorylated R-Smads then form heteromeric complexes with the co-Smad, Smad4 and translocate to the nucleus to regulate TGFβ responsive genes. Following nuclear translocation, heteromeric Smad complexes regulate TGFβ-responsive genes via interaction with specific promoter sequences, termed Smad binding elements (SBEs). Increased TGFβ signaling, also upregulates I-Smads, Smad6 and Smad7, which inhibit phosphorylation of R-Smads and target TGFβ receptors for degradation, thus forming a negative feedback loop for the TGFβ signaling pathway (Izzi and Attisano, 2004; Massague et al., 2005). Moreover, several phosphatases that dephosphorylate R-Smads to regulate TGFβ signaling were recently identified (Chen et al., 2006; Lin et al., 2006).

While only one TGFβRII has been identified in mammals, there are at least two different type one receptors (TGFβRIs). The classical TGFβRI, also known as activin receptor-like kinase (ALK)-5, is expressed on almost all cell types (Goumans et al., 2003; Wrana et al., 1994). In contrast, the other TGFβRI, ALK1, is preferentially expressed by endothelial cells. Activation of each TGFβRI utilizes different R-Smads and regulates expression of distinct sets of genes, leading to different cellular effects. When ALK5 is activated, it phosphorylates Smad2 and Smad3 and turns on the expression of genes including collagen I and plasminogen activator inhibitor I (PAI-1). In contrast, ALK1 activation phosphorylates Smad1, -5, and -8 to regulate expression of genes such as Id inhibitors of differentiation (Id1, -2, -3) (Goumans et al., 2002). Although Smads are critical for TGFβ signal transduction, compelling evidence has suggested that Smad-independent pathways may also mediate TGFβ signaling. For instance, TGFβ has been found to activate mitogen-activated protein kinases (MAPKs), phosphoinositide 3-kinase (PI3K), and Rho family guanosine triphosphatases (GTPases) to regulate cell growth and apoptosis (Massague et al., 2005).

TGFβ has dual effects on cell proliferation, differentiation, angiogenesis, and inflammation. For instance, TGFβ promotes proliferation of mesenchymal cells such as dermal fibroblasts, but it acts as a potent growth inhibitor of epithelial cells (e.g., keratinocytes), hematopoietic cells (e.g., lymphocytes), and neural cells by inducing anti-proliferative gene responses during cell division (Massague, 2000; Massague et al., 2000). In endothelia, ALK1-mediated TGFβ signaling leads to endothelial cell proliferation and migration that are essential for angiogenesis, whereas ALK5 activation opposes this effect, thereby preventing angiogenesis (Goumans et al., 2003). TGFβ is a strong chemoattractant for all leukocytes including granulocytes, lymphocytes, monocytes/macrophages, and mast cells, and is thus able to initiate inflammatory responses. However, TGFβ later deactivates leukocytes and contributes to the resolution of inflammation (Wahl, 1992; Wahl, 1994). Due to the cell type- and stage-specific roles of TGFβ, the role of TGFβ signaling in the maintenance of tissue homeostasis and in disease development is complex.

Alterations of TGFβ Pathway Components in Skin and Head-and-Neck Cancer

The TGFβ signaling pathway plays an important role in tumor suppression, primarily via growth inhibition, apoptosis, and maintenance of differentiation. Paradoxically, TGFβ signaling also promotes tumor progression and metastasis through increased angiogenesis, inflammation, and epithelial-mesenchymal transition (EMT) (Bierie and Moses, 2006). We have shown that human skin SCCs exhibit increased levels of TGFβ1, but reduced levels of TGFβRII protein (Li et al., 2005). Moreover, carcinoma *in situ* samples already exhibited these changes compared to normal epidermis (Li et al., 2005). Similarly, we have determined that TGFβ1 protein is overexpressed in HNSCC samples as well as mucosa adjacent to the tumor, whereas TGFβRII was lost at the mRNA and protein level in HNSCC only (Lu et al., 2004). Therefore, while TGFβ signaling through TGFβRII is preserved during cancer formation, aggressive HNSCC cells lose TGFβ1-mediated growth inhibition via loss of TGFβRII or other molecular alterations.

Spatial and Temporal Effects of TGFβ Overexpression on Carcinogenesis

TGFβ has been thought to inhibit proliferation and promote differentiation of keratinocytes. We have shown that in human skin cancers, TGFβ1 is overexpressed either suprabasally or throughout the tumor epithelia including basal proliferative cells (Li et al., 2005). Whether or not the spatial patterns of TGFβ1 expression in tumor epithelia affects tumor prognosis is currently unknown. However, our studies using various transgenic mouse models have revealed functional differences, particularly in keratinocyte proliferation and inflammation, which may differentially affect tumor outcome. TGFβ was initially documented as a potent growth inhibitory cytokine of keratinocytes *in vitro* (Pittelkow et al., 1988). Similarly, transgenic mice overexpressing a constitutively active form of TGFβ1 (TGFβ1^{act}) in the differentiated keratinocytes of the epidermis driven by a human K1 promoter (HK1.TGFβ1^{act}) showed significant growth arrest in the epidermis (Sellheyer et al., 1993). Using a gene-switch transgenic mouse model we have also shown that acute induction of TGFβ1^{act} overexpression primarily in suprabasal layers of the epidermis reduces epidermal proliferation in quiescent skin or causes a resistance to phorbol 12-myristate 13-acetate (TPA)-induced epidermal hyperproliferation (Wang et al., 1999). Sustained induction of suprabasal TGFβ1^{act} overexpression achieved by application of the transgene expression inducer to these mice daily for 7 days revealed no overt changes in the epidermis but significantly increased angiogenesis in the dermis (Wang et al., 1999). In contrast, constitutive overexpression of latent TGFβ1 (TGFβ1^{wt}) (Li et al., 2004b) or chronically induced overexpression of TGFβ1^{act} in the basal layer (Liu et al., 2001) under the control of a K5 promoter led to epidermal hyperproliferation accompanied with inflammation. Lastly, studies from R. Akhust laboratory have demonstrated that when TGFβ1^{act} transgene is constitutively overexpressed in the suprabasal layers of the epidermis driven by a truncated human K10 promoter (Cui et al., 1995) or inducibly overexpressed by TPA under the control of a truncated K6 promoter (Fowlis et al., 1996), transgenic mice show an increased epidermal proliferative rate without histological changes. No obvious phenotypes in the dermis have been reported from these mice. These contradictory data have suggested that the effects of TGFβ1 overexpression on both the epidermis and the dermis may vary depending on expression level and spatial pattern (suprabasal or basal).

When subjected to skin chemical carcinogenesis protocol, transgenic mice overexpressing TGFβ1^{act} in the suprabasal layers of the epidermis show reduced papilloma formation, yet enhanced malignant conversion (Cui et al., 1996). This has led to a well-accepted concept that TGFβ inhibits benign tumor formation at early stages of skin carcinogenesis, but enhances malignant progression at later stages (Cui et al., 1996). In agreement with the tumor promotion role of TGFβ at later stages during skin carcinogenesis, we have demonstrated that directly inducing TGFβ1 overexpression in suprabasal layers of papilloma epithelia of gene-switch-TGFβ1 transgenic mice causes accelerated malignant transformation and rapid metastasis (Li et al., 2005; Weeks et al., 2001). This is associated with an upregulation of MMP genes and an increased angiogenesis, indicative of a paracrine effects of TGFβ1 on stromal cells such as fibroblasts and inflammatory cells (Li et al., 2005; Weeks et al., 2001).

Role of TGFβ in Inflammation and Carcinogenesis

Inflammation has been associated with cancer development, and many inflammatory cytokines and chemokines stimulate cell proliferation and angiogenesis, which facilitate tumor growth (Coussens and Werb, 2002; Vicari and Caux, 2002). With respect to skin cancer, paradoxical effects of inflammation on cancer development have also been reported. Psoriatic patients with chronic skin inflammation do not show an increased risk of developing skin cancer on psoriatic plaques (Nickoloff, 2004). However, in a human papillomavirus oncogene-induced skin carcinogenesis model, mice devoid of pro-inflammatory CD4+ T cells exhibited a lower incidence of tumors and delayed neoplastic progression, providing an unexpected role of CD4+ T cells in immune enhancement of skin carcinogenesis (Daniel et al., 2003). In order to assess the role of TGFβ1 in skin inflammation and carcinogenesis, we generated transgenic mice expressing wild-type TGFβ1 in basal keratinocytes and hair follicles (K5.TGFβ1^{wt}). Transgenic mice developed a severe inflammatory skin disorder (Li et al., 2004b). Interestingly, histological examination of K5.TGFβ1^{wt} transgenic skin demonstrated inflammatory cell infiltration and angiogenesis as early as day 17 after birth, when the K5.TGFβ1^{wt} mice were macroscopically indistinguishable from non-transgenic littermates (Li et al., 2005; Li et al., 2004b), suggesting that epidermal hyperplasia may be a secondary effect of inflammation and angiogenesis. Molecular analysis revealed that the skin inflammatory phenotype was associated with an upregulation in a variety of genes encoding inflammatory cytokines [*e.g.*, IL-1, IL-2, TNFα, and interferon (IFN)-γ], chemokines [*e.g.*, MIP-2 (murine counterpart of IL-8), monocyte-chemotactic protein (MCP)-1, and interferon-induced protein-10 (IP-10)], growth regulators [*e.g.*, insulin-like growth factor (IGF)-1, amphiregulin, and KGF], matrix metalloproteinases (MMPs) (*e.g.*, MMP-2, -3, and –9), and angiogenic factors (*e.g.*, VEGF) and their receptors (*e.g.*, Flt-1 and Flk-1) (Li et al., 2004b). These molecules participate in epidermal proliferation, inflammatory cell infiltration, angiogenesis, and basement membrane degradation, all of which play a part in skin carcinogenesis.

To further assess the role of TGFβ1-associated inflammation in carcinogenesis, we turned our attention to HNSCCs. In comparison with skin SCCs, HNSCCs more frequently develop from sites of chronic inflammation (*e.g.*, oral lichen planus) and have a much worse prognosis (Coussens and Werb, 2002; Philpott and Ferguson, 2004). To further determine whether early stage TGFβ1 overexpression directly induces inflammation, we induced TGFβ1 transgene expression in the basal layer of head and neck epithelia in our keratinocyte-specific gene-switch transgenic mice. Similar to the effect of TGFβ1 overexpression in the epidermis (Li et al., 2004b), we found that TGFβ1 transgene induction in the oral mucosa also resulted in inflammation, increased angiogenesis, and subsequent epithelial hyperproliferation (Lu et al., 2004). Given the importance of inflammation in cancer development, our data suggest that in certain tissue types, TGFβ1-induced inflammation may override its tumor suppressive effect even at early stages of carcinogenesis.

This notion is further supported by our recent finding that deletion TGFβRII in head and neck epithelia resulted in an elevated endogenous TGFβ1, which was associated with severe inflammation and angiogenesis in head and neck tissue (Lu et al., 2006). In combination with an activated Ras oncogene in TGFβRII null epithelial cells, these mice developed HNSCCs that mimicked invasive human HNSCCs at the pathological and molecular levels.

Role of TGFβ Signaling in EMT and Metastasis

To assess the effect of combined TGFβ1 overexpression and loss of TGFβRII in skin cancer, we generated transgenic mice that allowed inducible expression of TGFβ1 in keratinocytes expressing a dominant negative TGFβRII (ΔβRII) in the epidermis (Han et al., 2005; Li et al., 2005). Induction of TGFβ1 transgene expression alone in late stage chemically-induced papillomas failed to inhibit tumor growth but increased metastasis and epithelial-to-mesenchymal transition (EMT), i.e., formation of spindle cell carcinomas. Although ΔβRII expression in tumor epithelia was able to abrogate TGFβ1-mediated EMT, a well known metastasis-associated oncogenic event, it promoted tumor metastasis in cooperation with TGFβ1. TGFβ1/ΔβRII-transgenic tumors progressed to metastasis without losing membrane-associated E-cadherin/catenin complexes, and at a rate higher than those observed in non-transgenic, TGFβ1 transgenic, or ΔβRII transgenic mice. Abrogation of Smad activation by ΔβRII correlated with the blockade of EMT. However, ΔβRII did not alter TGFβ1-mediated expression of RhoA/Rac, Erk or JNK, which contributed to increased metastasis (Li et al., 2005). Our study provides evidence that TGFβ1 induces EMT and invasion via distinct mechanisms. TGFβ1-mediated EMT requires functional TGFβRII, whereas TGFβ1-mediated tumor invasion cooperates with reduced TGFβRII signaling in tumor epithelia. The present study has raised a very interesting conclusion that TGFβ1-mediated EMT in tumor epithelia is dispensable for tumor invasion and metastasis, which can be largely contributed to by the effects of TGFβ1 on the stroma, where TGFβ responsiveness remains intact. Thus, TGFβ1-mediated tumor invasion and EMT can be uncoupled.

To determine whether downstream TGFβ signaling components are important for SCC formation and EMT, we deleted individual Smads in murine keratinocytes. Deletion of Smad4 in the epidermis resulted in spontaneous SCC formation (Qiao et al., 2006). However, no spindle cell carcinomas resulting from EMT were observed. Interestingly, increased nuclear Smad3 levels were observed in Smad4 knockout SCCs (unpublished data). This result suggests that Smad4 loss abrogates TGFβ-mediated EMT but not TGFβ–mediated invasion, the latter of which may be mediated by Smad3.

While there are numerous reports of Smad4 loss and/or mutation in various cancer types (Bierie and Moses, 2006), Smad3 loss or inactivation is very rare (Sjoblom et al., 2006). We have found that Smad3-/- and Smad3+/- mice are resistant to two-stage skin chemical carcinogenesis (Li et al., 2004a). Moreover, we found a dramatic reduction in inflammation in Smad3 knockout tumors compared to wild-type tumors, with particular loss of tumor-associated macrophages (TAMs) (Li et al., 2004a). It is well known that TPA treatment induces skin inflammation, and this stromal effect is critical for cancer development. TGFβ1 is a major cytokine induced by TPA. We have shown that TGFβ1 overexpression in keratinocytes has a profound effect on inducing skin inflammation (Li et al., 2004b). Therefore, our study suggests that Smad3 is required for TGFβ1-mediated inflammation. Additionally, Smad3 knockout skin demonstrated reduced epidermal hyperplasia, reduced proliferation in papillomas, and increased apoptosis in skin and papillomas compared to wild-type mice. TPA-induction increases TGF-β1 levels in keratinocytes, which normally leads to the up-regulation of tumor promoting factors, including TGF-β and AP-1 family members such as c-fos, c-jun, junB, and junD. However, Smad3 knockout TPA-treated skin, papillomas, and tumors demonstrated decreased induction of these molecules (Li et al., 2004a), indicating that Smad3 is required for this aspect of TPA-related tumor promotion.

Conclusions

Our studies suggest that TGFβ1-mediated inflammation plays an important role in tumor promotion. SCCs with TGFβ1 overexpression can still metastasize without EMT through increased inflammation, protease activity, and angiogenesis. Thus, SCCs that have both increased TGFβ1 and reduced TGFβRII or Smad4 in tumor epithelia will have a poorer prognosis than those with TGFβ1 overexpression alone or loss of TGFβRII or Smad4 alone. Taken together, inhibition of the remaining TGFβ1 effect on tumor stroma in combination with targeting oncogenic changes in cancer epithelia may provide an effective therapy for SCC.

References

Bierie, B., and Moses, H. L. (2006). TGF-beta and cancer. *Cytokine Growth Factor Rev* 17, 29-40.

Chen, H. B., Shen, J., Ip, Y. T., and Xu, L. (2006). Identification of phosphatases for Smad in the BMP/DPP pathway. *Genes Dev* 20, 648-653.

Coussens, L. M., and Werb, Z. (2002). Inflammation and cancer. *Nature* 420, 860-867.

Cui, W., Fowlis, D. J., Bryson, S., Duffie, E., Ireland, H., Balmain, A., and Akhurst, R. J. (1996). TGFbeta1 inhibits the formation of benign skin tumors, but enhances progression to invasive spindle carcinomas in transgenic mice. *Cell* 86, 531-542.

Cui, W., Fowlis, D. J., Cousins, F. M., Duffie, E., Bryson, S., Balmain, A., and Akhurst, R. J. (1995). Concerted action of TGF-beta 1 and its type II receptor in control of epidermal homeostasis in transgenic mice. *Genes Dev* 9, 945-955.

Daniel, D., Meyer-Morse, N., Bergsland, E. K., Dehne, K., Coussens, L. M., and Hanahan, D. (2003). Immune enhancement of skin carcinogenesis by CD4+ T cells. *J Exp Med* 197, 1017-1028.

Feng, X. H., and Derynck, R. (2005). Specificity and versatility in tgf-beta signaling through Smads. *Annu Rev Cell Dev Biol* 21, 659-693.

Fowlis, D. J., Cui, W., Johnson, S. A., Balmain, A., and Akhurst, R. J. (1996). Altered epidermal cell growth control in vivo by inducible expression of transforming growth factor beta 1 in the skin of transgenic mice. *Cell Growth Differ* 7, 679-687

Goumans, M. J., Lebrin, F., and Valdimarsdottir, G. (2003). Controlling the angiogenic switch: a balance between two distinct TGF-b receptor signaling pathways. *Trends Cardiovasc Med* 13, 301-307.

Goumans, M. J., Valdimarsdottir, G., Itoh, S., Rosendahl, A., Sideras, P., and ten Dijke, P. (2002). Balancing the activation state of the endothelium via two distinct TGF-beta type I receptors. *Embo J* 21, 1743-1753.

Han, G., Lu, S. L., Li, A. G., He, W., Corless, C. L., Kulesz-Martin, M., and Wang, X. J. (2005). Distinct mechanisms of TGF-beta1-mediated epithelial-to-mesenchymal transition and metastasis during skin carcinogenesis. *J Clin Invest* 115, 1714-172

Izzi, L., and Attisano, L. (2004). Regulation of the TGFbeta signalling pathway by ubiquitin-mediated degradation. *Oncogene* 23, 2071-2078.

Li, A. G., Koster, M. I., and Wang, X. J. (2003). Roles of TGFbeta signaling in epidermal/appendage development. *Cytokine Growth Factor Rev* 14, 99-111.

Li, A. G., Lu, S. L., Han, G., Kulesz-Martin, M., and Wang, X. J. (2005). Current view of the role of transforming growth factor beta 1 in skin carcinogenesis. *J Investig Dermatol Symp Proc* 10, 110-117.

Li, A. G., Lu, S. L., Zhang, M. X., Deng, C., and Wang, X. J. (2004a). Smad knockout mice exhibit a resistance to skin chemical carcinogenesis. *Cancer Re* 64, 7836-7845.

Li, A. G., Wang, D., Feng, X. H., and Wang, X. J. (2004b). Latent TGFbeta overexpression in keratinocytes results in a severe psoriasis-like skin disorder *Embo J* 23, 1770-1781.

Lin, X., Duan, X., Liang, Y. Y., Su, Y., Wrighton, K. H., Long, J., Hu, M., Davis, C M., Wang, J., Brunicardi, F. C., *et al.* (2006). PPM1A functions as a Sma phosphatase to terminate TGFbeta signaling. *Cell* 125, 915-928.

Liu, X., Alexander, V., Vijayachandra, K., Bhogte, E., Diamond, I., and Glick, A (2001). Conditional epidermal expression of TGFbeta 1 blocks neonatal lethalit but causes a reversible hyperplasia and alopecia. *Proc Natl Acad Sci U S A* 98 9139-9144.

Lu, S. L., Herrington, H., Reh, D., Weber, S., Bornstein, S., Wang, D., Li, A. G., Tang, C. F., Siddiqui, Y., Nord, J., *et al.* (2006). Loss of transforming growth factor-beta type II receptor promotes metastatic head-and-neck squamous cell carcinoma. *Genes Dev* 20, 1331-1342.

Lu, S. L., Reh, D., Li, A. G., Woods, J., Corless, C. L., Kulesz-Martin, M., and Wang, X. J. (2004). Overexpression of transforming growth factor beta1 in head and neck epithelia results in inflammation, angiogenesis, and epithelial hyperproliferation. *Cancer Res* 64, 4405-4410.

Massague, J. (2000). How cells read TGF-beta signals. *Nat Rev Mol Cell Biol* 1, 169-178.

Massague, J., Blain, S. W., and Lo, R. S. (2000). TGFbeta signaling in growth control, cancer, and heritable disorders. *Cell* 103, 295-309.

Massague, J., and Gomis, R. R. (2006). The logic of TGFbeta signaling. *FEBS Lett* 580, 2811-2820.

Massague, J., Seoane, J., and Wotton, D. (2005). Smad transcription factors. *Genes Dev* 19, 2783-2810.

Nickoloff, B. J. (2004). The skin cancer paradox of psoriasis: a matter of life and death decisions in the epidermis. *Arch Dermatol* 140, 873-875.

Philpott, M., and Ferguson, L. R. (2004). Immunonutrition and cancer. *Mutat Res* 551, 29-42.

Pittelkow, M. R., Coffey, R. J., Jr., and Moses, H. J. (1988). Keratinocytes produce and are regulated by transforming growth factors. *Ann N Y Acad Sci* 548, 211-224.

Qiao, W., Li, A. G., Owens, P., Xu, X., Wang, X. J., and Deng, C. X. (2006). Hair follicle defects and squamous cell carcinoma formation in Smad4 conditional knockout mouse skin. *Oncogene* 25, 207-217.

Sellheyer, K., Bickenbach, J. R., Rothnagel, J. A., Bundman, D., Longley, M. A., Krieg, T., Roche, N. S., Roberts, A. B., and Roop, D. R. (1993). Inhibition of skin development by overexpression of transforming growth factor beta 1 in the epidermis of transgenic mice. *Proc Natl Acad Sci U S A* 90, 5237-5241.

Sjoblom, T., Jones, S., Wood, L. D., Parsons, D. W., Lin, J., Barber, T. D., Mandelker, D., Leary, R. J., Ptak, J., Silliman, N., *et al.* (2006). The consensus coding sequences of human breast and colorectal cancers. *Science* 314, 268-274.

Vicari, A. P., and Caux, C. (2002). Chemokines in cancer. *Cytokine Growth Factor Rev* 13, 143-154.

Wahl, S. M. (1992). Transforming growth factor beta (TGF-beta) in inflammation: a cause and a cure. *J Clin Immunol* 12, 61-74.

Wahl, S. M. (1994). Transforming growth factor beta: the good, the bad, and the ugly. *J Exp Med* 180, 1587-1590.

Wang, X. J. (2001). Role of TGFbeta signaling in skin carcinogenesis. *Microsc Res Tech* 52, 420-429.

Wang, X. J., Liefer, K. M., Tsai, S., O'Malley, B. W., and Roop, D. R. (1999). Development of gene-switch transgenic mice that inducibly express transforming growth factor beta1 in the epidermis. *Proc Natl Acad Sci U S A* 96, 8483-8488.

Weeks, B. H., He, W., Olson, K. L., and Wang, X. J. (2001). Inducible expression of transforming growth factor beta1 in papillomas causes rapid metastasis. *Cancer Res* 61, 7435-7443.

Wrana, J. L., Attisano, L., Wieser, R., Ventura, F., and Massague, J. (1994). Mechanism of activation of the TGF-beta receptor. *Nature* 370, 341-347.

TGF-β and Tumor Progression Mediated by Reactive Oxygen Species

Kiyoshi Nose, Tadashi Sugimoto, Masami Wakamatsu, Fumihiro Ishikawa and Motoko Shibanuma

Department of Microbiology, Showa University School of Pharmaceutical Sciences, Hatanodai 1-5-8, Shinagawa-ku, Tokyo 142-8555, Japan

Summary. TGF-β1 is a multi-functional cytokine. It suppresses growth of normal epithelial cells, whereas it promotes malignant progression in preneoplastic cells. As signaling molecules of TGF-β, Smads are well known regulators, but we previously showed that TGF-β stimulated cells to increase intracellular ROS levels that participate in the expressions of several genes. ROS have recently been recognized as second messengers for physiological signaling. and we confirmed that reactive oxygen species (ROS) levels were elevated in TGF-β-treated cells. The mitochondria were the major source of ROS, and ROS participated in the induction of several TGF-β-inducible genes such as MMP-9, Nox4 and MMP-10. These molecules could be involved in the malignant progression of premalignant cells, and the activation of the mitochondria is an important step for this process.

Key words. TGF-β, reactive oxygen species, mitochondria, MMP-9

Results and Discussion

1 Production of reactive oxygen species

The intracellular (cytosolic) hydrogen peroxide level was monitored in a single cell with a fluorescent reporter (Hyper)-C (cytosol) and found to increase in TGF-β1-treated mouse mammary epithelial cells (NMuMG) during the first several hours after treatment (Fig. 1). The expression of a dominant negative form of ALK5 (type I receptor for TGF-β) attenuated the increase, indicating that the change in hydrogen peroxide level was downstream of the TGF-β receptor. Intriguingly, this increase was abolished in mitochondria-depleted (ρ0) cells. These observations suggest that TGF-β1 modify respiratory activity of mitochondria, thereby causing redistribution of hydrogen peroxide within cells. Consistent with this, the membrane potential of mitochondria was found to decrease by TGF-β1.

The smad-signal pathways seemed to be intact in rho-0 cells since phosphorylation of Smad3 and p38 protein was increased by TGF-β1 even in rho-0 cells.

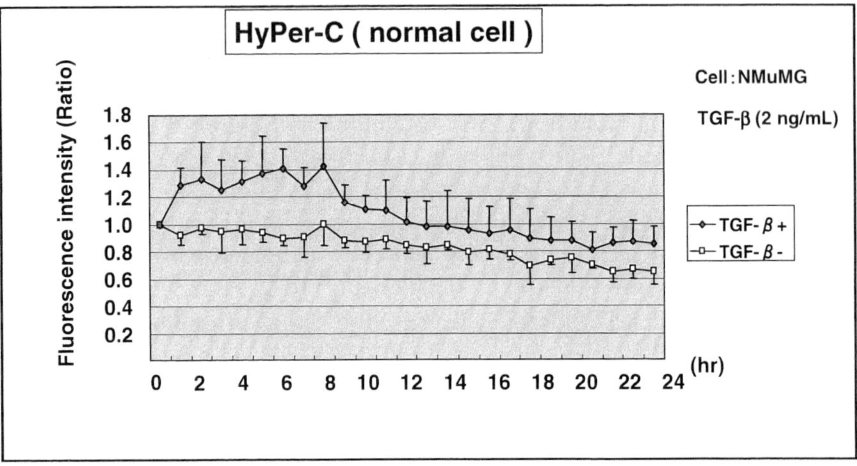

Fig. 1 Levels of hydrogen peroxide in control and TGF-β –treated cells. NMuMG cells were transfected with an expression vector for Hyper-C, and 16 – 18 hrs later cells were either untreated or treated with TGF-β. Fluorescent intensities of Hyper-C were measured under fluorescent microscope.

2 Mitochondria-dependent expression of genes

To address the significance of the above observation, we examined gene expression profiles regulated by TGF-β1 and their dependence on mitochondria. With gene tip analysis, several genes were identified that were induced by TGF-β1 in a mitochondria-dependent manner. Among them, cell adhesion- as well as mesenchymal phenotype-related genes were predominantly included (Table 1). It is, thus, speculated that TGF-β1 modulates mitochondrial respiratory activity to redistribute reactive oxygen species in cells, and ROS in turn function as mediators regulating the gene expressions that are critical for malignant phenotypes such as invasion, cell movement and adhesion.

Some examples of candidate genes such as MMP-9, integrin α11, smooth muscle actin, adam12 and surpin f1 were selected, and inducibility was confirmed by RT-PCR. The induction of these genes by TGF-β1 was decreased in ρ0 cells, whereas that of fibronectin, integrin α5, and PAI-1 genes were not significantly affected.

TGF-β -inducible and mitochondria-dependent genes ~ DNA microarray analysis ~

OGS	T(+/-)	T(+/-)ρ	description			
Itga11	44.81	3.03	integrin, alpha 11 (Itga11)	○	△	★
Itga5	26.94	4.03	integrin alpha 5 (fibronectin receptor alpha) (Itga5)	○		★
Ctsw	25.03	11.33	cathepsin W (Ctsw)			
Ltbp2	23.7	10.61	procollagen, type V, alpha 3 (Col5a3)		△	
Col5a3	23.2	5.5	latent transforming growth factor beta binding protein 2 (Ltbp2)	○	△	
Serpinf1	22.5	4	serine (or cysteine) proteinase inhibitor, clade F, member 1 (Serpinf1)		△	★
Acta2	21.57	2.04	actin, alpha 2, smooth muscle, aorta (Acta2)	○	△	★
Col7a1	17.54	6.24	procollagen, type VII, alpha 1 (Col7a1)	○		
Acta2	14.34	1.87	actin, alpha 2, smooth muscle, aorta (Acta2)	○	△	
Eno2	13.72	-2.79	enolase 2, gamma neuronal (Eno2)			
Myl7	13	4.52	myosin, light polypeptide 7, regulatory (Myl7)	○		
Plk4	11.44	-1.18	polo-like kinase 4 (Drosophila) (Plk4)			
Hck	10.5	4.49	hemopoietic cell kinase (Hck)			
Tnc	9.59	4.56	tenascin C (Tnc)	○		
Acta1	6.73	3.27	actin, alpha 1, skeletal muscle (Acta1)	○		
Adam12	5.54	2.28	a disintegrin and metalloproteinase domain 12 (meltrin alpha) (Adam12)	○	△	★

○ Cell adhesion, cytoskeleton △ Mesenchyme (bone, muscle, cartilage, tendon etc.) -related
★ Analyzed in detail

Table 1. DNA microarray analysis of gene expression pattern. mRNA from control, TGF-β -treated or mitochondria-depleted NMuMG cells were analyzed by microarray. Genes whose expressions were modulated commonly by TGF-β-treatment or mitochondria-depletion were summarized

3 Induction of MMP-9

The induction of MMP-9 was significantly decreased in the mitochondria-depleted cells; this, however, required the TGF-β-receptor since the kinase-dead (KD) form of ALK5 decreased the induction. On the other hand, the induction was independent on the Smad pathway because the expression of the dominant negative form of Smad3 or inhibitory Smad (Smad 7) did not affect the induction. On the other hand, induction of fibronectin, integrin α5 and PAI-I was not affected by the mitochondrial depletion but was decreased by the expression of Smad7. These results indicate that signaling pathways for MMP-9 are partly different from those of FN, integrin a5 and PAI-I (Fig. 2).

MMP-9 induction was significantly decreased in the presence of radical scavengers such as tiron or PDTC, but induction of integrin α5 was not affected. As for the mechanism of MMP-9 induction, the induction seems to be transcriptional because levels of RNA containing introns, as well as exons, were increased by TGF-β.

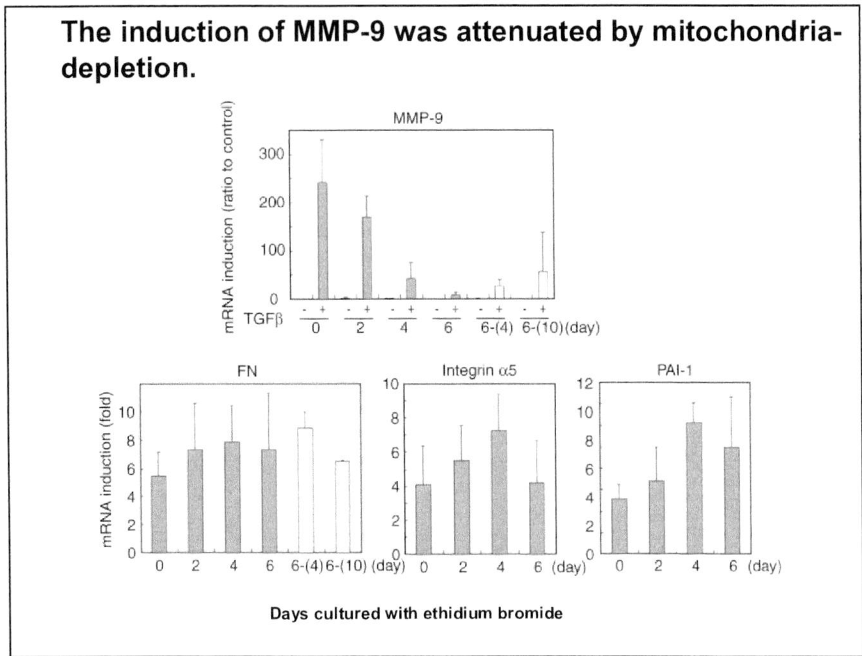

Fig. 2. Effect of mitochondria-depletion on gene expressions induced by TGF-β. NMuMG cells were either untreated or treated with ethidium bromide for indicated times to deplete mitochondria, and were treated with TGF-β. Levels of mRNA for MMP-9, fibronectin (FN), integrin α5, or plasminogen activator inhibitor (PAI)1 were measured by quantitative RT-PCR.

4 Induction of Nox4

The next question was concerned with the involvement of NADPH oxidase in hydrogen peroxide production in TGF-β-treated cells. NADPH oxidase is known to form a family that contains 5 or more members, and each member is comprised of catalytic and regulatory subunits. NMuMG cells express Nox4, and the mRNA levels of this protein were examined.

Nox4 and its regulatory subunit increased dramatically in TGF-β1-treated cells both in terms of mRNA and protein, but Nox2 or gp91 mRNA levels decreased 24 hr following TGF-β1-addition. The cells exposed to EtBr for 2, 4, 6 days and those further cultured for 4 days without EtBr were treated with TGF-β1. Nox4 induction was decreased in mitochondria-depleted cells.

The effect of inhibitors for ROS production and ROS scavengers on the induction of Nox4 by TGF-β was examined. DPI, that is a nonspecific inhibitor of ROS production, and rotenone significantly inhibited Nox4 induction; scavengers for ROS or superoxide also inhibited Nox4 induction.

The involvement of Nox-4 in TGF-β-signaling was examined by knockdown using SiRNA. When SiRNA for Nox4 was introduced to the cells, induction of MMP-10 was inhibited, whereas that of other genes was not affected significantly. These results indicate that Nox4 is important for the induction of MMP-10 possibly through the production of ROS.

In summary, H2O2 produced from the TGF-β-treated cells participates in the regulation of the expressions of several genes that work as tumor progressors.

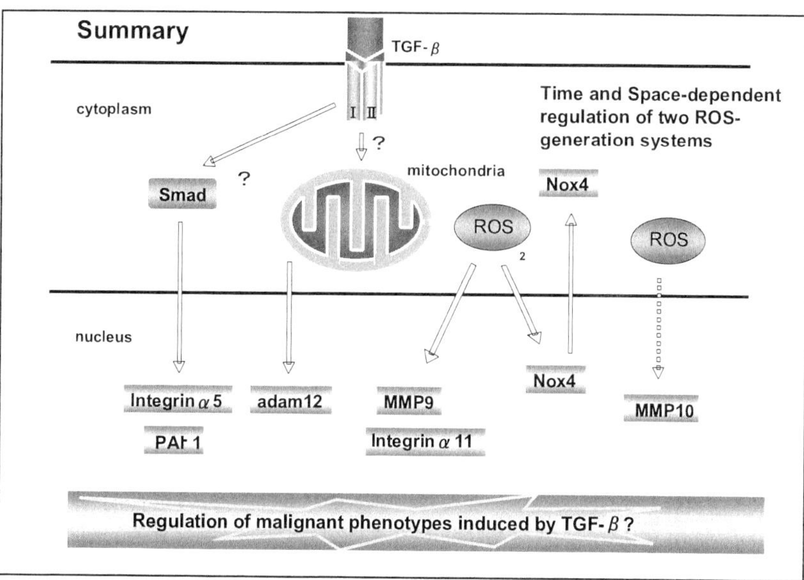

Fig. 3 Summary Refer to color plates.

References

Kanome T, Itoh N, Mori K, Kim-Kaneyama J, Shibanuma M, Nose K. Characterization of Jumping translocation breakpoint (JTB) gene product isolated as a TGF-β1-inducible clone involved in regulation of mitochondrial function, cell growth and death. Oncogene (in press).

Ohba M, Shibanuma M, Kuroki T, Nose K. (1994)Production of hydrogen peroxide by transforming growth factor β1 and its involvement in induction of egr-1 in mouse osteoblastic cells. J Cell Biol 126: 1079-1088

Role of MT1-MMP in Tumor-Stromal Interaction

Motoharu Seiki

Division of Cancer Cell Research, Institute of Medical Science, University of Tokyo, Minato-ku, Tokyo, 108-8639, Japan

Summary. MMP-2 is a stroma-derived matrix metalloproteinase and also known as a type IV collagenase. It is believed to mediate tumor cell behavior by degrading deposits of type IV collagen, a major component of the basement membrane. The membrane-type matrix metalloproteinase MT1-MMP is a highly potent activator of MMP-2, and is expressed in many tumor and stromal cells. However, the roles played by stromal MMP-2 and tumor-derived MT1-MMP in tumor progression *in vivo* remain poorly understood. Thus, we analyzed importance of MT1-MMP/MMP-2 axis in vivo by using a colon epithelial cell line from an *Mt1-mmp*$^{-/-}$ mouse strain and *Mmp-2*$^{-/-}$ mice. Tumor formation of the cells was mostly dependent on expression of MT1-MMP in the cells and it additionally required MMP-2 supplied by the host tissue. On the other hand, overexpression of MMP-2 is frequent in tumor stroma and we identified EMMPRIN, a multifunctional glycoprotein, as a tumor-derived mediator to induce MMP-2 expression in fibroblasts after cleaved by MT1-MMP. MT1-MMP cleaves EMMPRIN and released a 22 kDa fragment which has activity to augment expression of MMP-2 in fibropbrasts. Thus, there is a positive feedback loop between the MT1-MMP/MMP-2-dependent tumor growth and stimulation of MMP-2 production in tumor stroma by MT1-MMP and EMMPRIN.

Key words. metalloproteinases, MMP-2, MT1-MMP, EMMPRIN, tumorigenicity

1. Introduction

Malignant tumors express multiple matrix metalloproteinases (MMPs), most of which are produced by stroma cells such as fibroblasts and infiltrating macrophages (Egeblad and Werb, 2002). MMPs are believed to play roles in tumor growth, invasion, and metastasis by degrading extracellular matrix (ECM) and by proteolytic conversion of bioactive molecules in tumor tissue (Egeblad and Werb, 2002). Based on this idea, many MMP inhibitors (MMPIs) have been developed for therapeutic use, although most clinical trials have ended in failure (Overall and Kleifeld,

2006). However, it should be noted that broad-spectrum MMPIs were used in these clinical studies, without precise knowledge about each MMP that should have been targeted. Thus, it is important to understand the functions of individual MMPs in tumor progression and we have been interested in roles of tumor-derived MT1-MMP and stroma-derived MMP-2 in tumor growth and invasion.

MMP-2 (matrix metalloproteinase 2; also known as gelatinase A and 72 kDa type IV collagenase) has the ability to degrade type IV collagen in the basement membrane (BM); it is one of the major stroma-derived MMPs (Sato et al., 1992). In spite of the many *in vitro* studies that have indicated that MMP-2 plays an important role in tumor invasion of the BM (Stetler-Stevenson, 1990), little is known about the specific functions of MMP-2 *in vivo*. MMP-2 is produced as an inactive zymogen (proMMP-2), which is then activated in a tumor-dependent manner (Stetler-Stevenson, 1990). The activation is mostly mediated by MT1-MMP (membrane-type 1 matrix metalloproteinase; also known as MMP-14) which is expressed in many tumor tissues (Seiki, 2003). In these tissues, MT1-MMP expression levels correlate well with MMP-2 activation, as well as a poor prognosis for the patient (Itoh and Seiki, 2006). Thus, MT1-MMP is believed to be a major activator of stroma-derived MMP-2 in tumors.

MT1-MMP is an integral membrane metalloproteinase that is expressed frequently both in tumor and stroma cells (Sato et al., 1994; Seiki, 2003). It cleaves a variety of ECM proteins (collagen I, II, III; fibronectin; laminin-5), cell adhesion molecules (CD44, integrin αv and tissue transglutaminase), cytokines and proMMPs (proMMP-2 and proMMP-13) (Itoh and Seiki, 2006). Among these substrates, type I collagen is a particularly important physiological target for MT1-MMP. Specifically, MT1-MMP-null mice are defective in type I collagen turnover and display severe fibrosis in their joints, delayed bone formation, multiple tissue defects and early death (Holmbeck et al., 1999). MT1-MMP is required *in vitro* for the invasion of type I collagen matrix by tumor and endothelial cells (Hotary et al., 2000). In addition, Hotary *et al.* (2003) reported that MT1-MMP is a critical requirement for tumor formation *in vivo*, and that *in vitro* its activity is sufficient for growth promotion of some tumor cell lines in type I collagen matrix (Hotary et al., 2003). Thus, it seems clear that, at least in cells surrounded by type I collagen, MT1-MMP type I collagenase activity is essential for tumor growth.

2. MT1-MMP is the major activator of MMP-2 in cancer

As proMMP-2 activation is believed to be an important step for cancer cells to invade into basal lamina, the mechanism of activation has been extensively studied. This process is not a simple interaction of proMMP-2 and MT1-MMP, but involves its endogenous inhibitor TIMP-2 (Seiki, 2003). MT1-MMP expressed on the cell surface forms a complex with TIMP-2 through the catalytic domain of the enzyme and the N-terminal inhibitory domain of TIMP-2 as an enzyme-inhibitor complex. The exposed C-terminal domain of TIMP-2 has an affinity for the hemopexine-like (Hpx) domain of proMMP-2, and this results in the formation of an MT1-MMP-TIMP-2-proMMP-2 ternary complex. Since MT1-MMP in this complex is inhibited by TIMP-2, another MT1-MMP free from TIMP-2, is required to carry out the activation of proMMP-2. To arrange another molecule of MT1-MMP next to the ternary complex of MT1-MMP-TIMP-2-proMMP-2, MT1-MMP forms a homo-oligomer complex through its Hpx domains and/or transmembrane/cytoplasmic domains (Itoh and Seiki, 2006). In this complex one of the MT1-MMP molecules acts as a receptor and the other acts as an activator, forming a proMMP-2 activation complex. This homo-oligomer complex formation is important in the activation process on the cell surface, because separating two MT1-MMPs by over expressing a catalytic domain deletion mutant of MT1-MMP effectively inhibits proMMP-2 activation (Itoh and Seiki, 2006).

3. EMMPRIN cleaved by MT1-MMP acts as an inducer of MMP-2 in stroma cells

EMMPRIN (extracellular matrix metalloproteinase inducer) is a multifunctional glycoprotein that belongs to the immunoglobulin superfamily (Biswas et al., 1995). EMMPRIN-null mice are sterile, and have defects in spermatogenesis, fertilization, sensory and memory functions, and mixed lymphocyte responses (Igakura et al., 1998). However, the exact mechanisms underlying the observed defects are still largely unknown.

EMMPRIN has been characterized to induce MMP-2, -3, -9, MT1-MMP, and MT2-MMP in addition to MMP-1; only glycosylated EMMPRIN is able to induce these MMPs, and the N-terminal Ig-like domain is also indispensable for the MMP-inducing activity, as well as for the homophilic interactions of the protein. EMMPRIN therefore potentially mediates the excessive production of MMPs in tumor tissue and is expected to act as a modulator of ECM in tumor tissues through the activity of the MMPs that it induces.

MT1-MMP (MMP-14) is an interesting candidate for the EMMPRIN shedding as an integral membrane protease responsible for pericellular proteolysis (Itoh and Seiki, 2006; Seiki, 2003). EMMPRIN shedding is enhanced by treatment of cells with phorbol-12-myristate 13-acetate

(PMA) and inhibited by tissue inhibitor of metalloproteinase (TIMP)-2 but not by TIMP-1, indicating the possible involvement of MT-MMPs. Indeed, expression, knockdown, and in vitro cleavage studies supported the involvement of MT1-MMP in EMMPRIN proteolysis (Egawa et al., 2006). The external portion of EMMPRIN is released from the cells by MT1-MMP cleavage and it retained the activity to induce expression of MMP-2 in fibroblasts. Thus, MT1-MMP expressed in tumor cells can increase MMP-2 in the tissue by releasing EMMPRIN fragment from the cells (Egawa et al., 2006).

4. Role of stroma-derived MMP-2 in MT1-MMP-dependent tumor growth

We established a colon epithelial cell line from an *Mt1-mmp*$^{-/-}$ mouse strain, and transfected these cells with an inducible expression system for MT1-MMP (MT1rev cells) (Taniwaki et al., 2007). Following subcutaneous implantation into *Mmp-2*$^{+/+}$ mice and induction of MT1-MMP expression, MT1rev cells grew rapidly, whereas they grew very slowly in *Mmp-2*$^{-/-}$ mice, even in the presence of MT1-MMP. This MT1-MMP-dependent tumor growth of MT1rev cells was enhanced in *Mmp-2*$^{-/-}$ mice as long as MMP-2 was supplied via transfection or co-implantation of MMP-2-positive fibroblasts. MT1rev cells cultured *in vitro* in a 3D collagen gel matrix also required the MT1-MMP/MMP-2 axis for rapid proliferation. MT1rev cells deposit type IV collagen primarily at the cell-collagen interface and these deposits appear scarce at sites of invasion and proliferation (Taniwaki et al., 2007). These data suggest that co-operation between stroma-derived MMP-2 and tumor-derived MT1-MMP may play a role in tumor invasion and proliferation via remodeling of the tumor-associated basement membrane.

5. Discussion

What clinical implications does the present study hold? Our findings suggest that MMP-2 plays a particularly important role in the early stages of tumors, when the BM remains adjacent to the tumor cells. At the same time, tumor cells in the early stages may retain the ability to express components of BM, including type IV collagen and laminin, as demonstrated here with MT1rev cells. On the other hand, the MMP-2 requirement may not be as critical for more advanced tumors, such as those that have completed the epithelial-mesenchymal transition (EMT). Even for advanced tumors, MT1-MMP may still be important since tumor cells are surrounded by type I collagen. Thus, MMP-2-targeted therapy may work effectively only in the early stages of tumors. However, it is noteworthy that the phenotypic characteristics of *Mmp-2*$^{-/-}$ mice are not severe (Itoh et al., 1997). Thus, specific inhibition of MMP-2 might provide effective prevention of the early stages of tumor progression, without causing serious side effects. However, as described earlier, the phenotypic characteristics of *Mt1-mmp*$^{-/-}$ mice are severe. Thus, only MT1-MMP expressed by tumor cells or in tumor tissue should be targeted if these severe side effects are to be avoided.

References

Biswas, C., Zhang, Y., DeCastro, R., Guo, H., Nakamura, T., Kataoka, H., and Nabeshima, K. (1995). The human tumor cell-derived collagenase stimulatory factor (renamed EMMPRIN) is a member of the immunoglobulin superfamily. Cancer Res *55*, 434-439.

Egawa, N., Koshikawa, N., Tomari, T., Nabeshima, K., Isobe, T., and Seiki, M. (2006). Membrane type 1 matrix metalloproteinase (MT1-MMP/MMP-14) cleaves and releases a 22-kDa extracellular matrix metalloproteinase inducer (EMMPRIN) fragment from tumor cells. J Biol Chem *281*, 37576-37585.

Egeblad, M., and Werb, Z. (2002). New functions for the matrix metalloproteinases in cancer progression. Nat Rev Cancer *2*, 161-174.

Holmbeck, K., Bianco, P., Caterina, J., Yamada, S., Kromer, M., Kuznetsov, S.A., Mankani, M., Robey, P.G., Poole, A.R., Pidoux, I., *et al.* (1999). MT1-MMP-deficient mice develop dwarfism, osteopenia, arthritis, and connective tissue disease due to inadequate collagen turnover. Cell *99*, 81-92.

Hotary, K., Allen, E., Punturieri, A., Yana, I., and Weiss, S.J. (2000). Regulation of cell invasion and morphogenesis in a three-dimensional type I collagen matrix by membrane-type matrix metalloproteinases 1, 2, and 3. J Cell Biol *149*, 1309-1323.

Hotary, K.B., Allen, E.D., Brooks, P.C., Datta, N.S., Long, M.W., and Weiss, S.J. (2003). Membrane type I matrix metalloproteinase usurps tumor growth control imposed by the three-dimensional extracellular matrix. Cell *114*, 33-45.

Igakura, T., Kadomatsu, K., Kaname, T., Muramatsu, H., Fan, Q.W., Miyauchi, T., Toyama, Y., Kuno, N., Yuasa, S., Takahashi, M., *et al.* (1998). A null mutation in basigin, an immunoglobulin superfamily member, indicates its important roles in peri-implantation development and spermatogenesis. Dev Biol *194*, 152-165.

Itoh, T., Ikeda, T., Gomi, H., Nakao, S., Suzuki, T., and Itohara, S. (1997). Unaltered secretion of beta-amyloid precursor protein in gelatinase A (matrix metalloproteinase 2)-deficient mice. J Biol Chem *272*, 22389-22392.

Itoh, Y., and Seiki, M. (2006). MT1-MMP: a potent modifier of pericellular microenvironment. J Cell Physiol *206*, 1-8.

Overall, C.M., and Kleifeld, O. (2006). Tumour microenvironment - opinion: validating matrix metalloproteinases as drug targets and anti-targets for cancer therapy. Nat Rev Cancer *6*, 227-239.

Sato, H., Kida, Y., Mai, M., Endo, Y., Sasaki, T., Tanaka, J., and Seiki, M. (1992). Expression of genes encoding type IV collagen-degrading metalloproteinases and tissue inhibitors of metalloproteinases in various human tumor cells. Oncogene *7*, 77-83.

Sato, H., Takino, T., Okada, Y., Cao, J., Shinagawa, A., Yamamoto, E., and Seiki, M. (1994). A matrix metalloproteinase expressed on the surface of invasive tumour cells. Nature *370*, 61-65.

Seiki, M. (2003). Membrane-type 1 matrix metalloproteinase: a key enzyme for tumor invasion. Cancer Lett *194*, 1-11.

Stetler-Stevenson, W.G. (1990). Type IV collagenases in tumor invasion and metastasis. Cancer Metastasis Rev *9*, 289-303.

Taniwaki, K., Fukamachi, H., Komori, K., Ohtake, Y., Nonaka, T., Sakamoto, T., Shiomi, T., Okada, Y., Itoh, T., Itohara, S., *et al.* (2007). Stroma-derived matrix metalloproteinase (MMP)-2 promotes membrane type 1-MMP-dependent tumor growth in mice. Cancer Res *67*, 4311-4319.

Role of Proinflammatory Prostaglandin E_2 in Bladder Tumor Progression

Shuntaro Hara

Department of Health Chemistry, School of Pharmaceutical Sciences, Showa University, 1-5-8 Hatanodai, Shinagawa-ku, Tokyo 142-8555, Japan

Summary. Cyclooxygenase-2 (COX-2), a rate-limiting enzyme in proinflammatory prostaglandin E_2 (PGE_2) biosynthesis, is up-regulated in a variety of human cancers, and multiple lines of evidence have suggested that COX-2 and COX-2-derived PGE_2 are important in carcinogenesis. Here, we found that COX-2 was markedly up-regulated in human bladder transitional cell carcinoma as well as in other malignancies. In normal bladder tissues, COX-2 protein was detected only in lympholid follicles. Expression of COX-2 in lymphoid follicles may be induced by proinflammatory stimuli and associated with bladder carcinogenesis. Furthermore, we found that among four types of PGE receptors, only EP1 was overexpressed in human bladder tumors and showed that EP1 may be involved in bladder tumor progression. Our results indicated that COX-2-derived PGE_2 promotes bladder carcinogenesis through the activation of EP1 receptor.

Keywords. prostaglandin E_2, cyclooxygenase-2, prostaglandin E receptor, bladder tumor

Introduction

Bladder cancer is the most common malignancy of the urinary tract, and the fourth or fifth leading cause of cancer-related death of men in Western industrialized countries (Jemal et al. 2007). The prognosis of patients with advanced bladder cancer is still extremely poor despite recent therapeutic advances, such as improved surgical techniques and perioperative combination chemotherapy. Therefore, future improvement in the survival rate of patients with bladder cancer might be possible through the development of novel indicator or therapeutic strategies.

Prostaglandin (PG) E_2, a proinflammatory lipid mediator, is involved in the carcinogenic process and malignant aggressiveness. Cyclooxygenase

(COX) is a crucial rate-limiting enzyme in PGE_2 biosynthesis, which converts arachidonic acid into PGG_2 and then PGH_2. Among two COX isozymes, COX-1 is constitutively expressed in nearly all tissues and functions in maintaining tissue homeostasis, whereas COX-2 is inducible by a large number of factors such as proinflammatory cytokines, growth factors, carcinogens, and tumor-promoting phorbol esters (Turini and DuBois 2002). Overexpression of COX-2 has been associated with the development of several types of cancer, including lung, colon, mammary and skin. Evidence from a variety of experimental approaches suggests that COX-2 and PGE_2 have procarcinogenic effects in these cancers by increasing cell proliferation, enhancing angiogenesis, promoting invasion, and inhibiting apoptosis (Wang et al. 2007). COX-2-derived PGE_2 exerts its biological actions through its binding to four specific receptor subtypes known as EP1, EP2, EP2 and EP4 (Sugimoto and Narumiya 2007). It has been reported that all of four EPs are involved in carcinogenesis.

Here, we revealed that COX-2 was markedly up-regulated in human bladder transitional cell carcinoma (TCC) as well as in other malignancies by RT-PCR and immunohistochemical analysis (Matsuzawa et al. 2002). It was also found that COX-2 immunostaining signals were observed in lymphoid follicles in human bladder tissues. Furthermore, we found that among four EPs, only EP1 was overexpressed in human bladder tumors and showed that EP1 may be involved in bladder tumor progression using mouse models.

Overexpression of COX-2 in human bladder cancer

A large body of evidence exists to suggest that COX-2 is important in gastroinstestinal cancer (Gupta and DuBois 2001). In order to determine whether COX-2 is overexpressed in TCC of the human bladder as well as in gastroinstestinal cancer, we investigated COX-2 expression in human TCC by RT-PCR and immunohistochemical analysis, and we found that normal bladder epithelium did not express COX-2 and that COX-2 was markedly up-regulated in human bladder TCC. Furthermore, the intensity and extent of COX-2 immunostaining in the bladder cancer tissues were scored and the relationship to tumor grade and stage was investigated. The levels of COX-2 expression were correlated with the tumor grade; from grade 1 to 3, there was stepwise increase in the COX-2 immunostaining score. These findings suggested that an increase in COX-2 expression may be associated with bladder carcinogenesis as well as gastrointestinal carcinogenesis, and that it may be useful as a biomarker in bladder cancer.

In normal tissues, COX-2 protein was detected only in lympholid follicles as well as in TCC. This observation is consistent with the previous reports that the expression of COX-2 is induced by proinflammatory stimuli and that inflammatory leukocytes express high levels of COX-2 protein (Nanayama et al. 1995, Niiro et al. 1997). It has also been reported that eosinophilic or irritant-induced cystitis is a risk factor for bladder cancer (Burin et al. 1995). Expression of COX-2 in lymphoid follicles may be induced by proinflammatory stimuli and associated with bladder carcinogenesis.

Involvement of EP1 in bladder tumor progression

We next investigated the expression of EP mRNAs in bladder tumor tissues by RT-PCR analysis and found that among four EPs, only EP1 was expressed. EP1 immunoreactive signals were also observed in human bladder TCC. Furthermore, our quantitative real-time RT-PCR analysis revealed that EP1 expression levels in bladder tumors were significantly increased in comparison to those in normal bladder tissues from the same patients.

Bladder carcinogenesis is associated with chemical exposure. BBN (*N*-n-butyl-*N*-butan-4-ol-nitrosoamine) is often used for promoting bladder carcinogenesis in mice. In order to reveal the involvement of EP1 in bladder carcinogenesis, we next investigated the effects of EP1 gene knockout and EP1 antagonist on BBN-induced bladder carcinogenesis in mouse models. As the results, the incidence of bladder tumors was significantly reduced by EP1 gene deficiency or administration of EP1 antagonist. The malignant potential of bladder tumors was also suppressed. These results indicated that COX-2-derived PGE_2 promotes bladder carcinogenesis through its binding to EP1 receptor.

Conclusion

We here revealed that both COX-2 and EP1 were overexpressed in human bladder tumors, and suggested that both of them may be useful as biomarkers in bladder cancer. Furthermore, our results indicated that COX-2-derived PGE_2 is involved in bladder tumor progression through the activation of EP1 receptor. Selective EP1 antagonists might be promising candidates for bladder cancer chemopreventive agents.

References

Burin GJ, Gibb HJ and Hill RN. (1995) Human bladder cancer: evidence for a potential irritation-induced mechanism. *Food Chem Toxicol* **33** 785-795

Gupta RA and DuBois RN. (2001) Colorectal cancer prevention and treatment by inhibition of cyclooxygenase-2. *Nat Rev Cancer* **1** 11-21

Jemal A, Siegel R, Ward E, Murray T, Xu J and Thun MJ. (2007) Cancer statistics, 2007. *CA Cancer J Clin* **57** 43-66

Matsuzawa I, Kondo Y, Kimura G, Hashimoto Y, Horie S, Imura N, Akimoto M and Hara S. (2002) Cyclooxygenase-2 expression and relationship to malignant potential in human bladder cancer. *J Health Sci* **48** 42-47

Nanayama T, Hara S, Inoue H, Yokoyama C and Tanabe T. (1995) Regulation of two isozymes of prostaglandin endoperoxide synthase and thromboxane synthase in human monoblastoid cell line U937. *Prostaglandins* **49** 371-382

Niiro H, Otsuka T, Izuhara K, Yamaoka K, Ohshima K, Tanabe T, Hara S, Nemoto Y, Tanaka Y, Nakashima H and Niho Y. (1997) Regulation by interleukin-10 and interleukin-4 of cyclooxygenase-2 expression in human neutrophils. *Blood* **89** 1621-1628

Sugimoto Y and Narumiya S. (2007) Prostaglandin E receptors. *J Biol Chem* **282** 11613-11617

Turini ME and DuBois RN. (2002) Cyclooxygenase-2: a therapeutic target. *Ann Rev Med* **53** 35-57

Wang MT, Honn KV and Nie D. (2007) Cyclooxygenases, prostanoids, and tumor progression. *Cancer Metastasis Rev* **26** 525-534

Part III

Biomarkers for Clinical Oncology

Molecular Biomarker for Lung Cancer

Hsuan-Yu Chen[1], Pan-Chyr Yang[2]

[1] Institute of Statistical Science, Academia Sinica, 128, Sec.2, Academia Road, Taipei 115, Taiwan
[2] Department of Internal Medicine, National Taiwan University Collage of Medicine, 1, Sec 1, Ren-Ai Rd, Taipei 100, Taiwan

Summary. Among lung cancer genres, Non-small cell lung cancer (NSCLC) is the predominant type and the most common cause of cancer deaths worldwide (Jemal et al. 2006, Pepe et al. 2004). A report from Taiwan Cancer Registry in 2002 demonstrated that the mortality of lung cancer is less than liver cancer in male, but it ranked number 1 in cancer mortality in female. Delayed diagnosis and poor treatment outcome are the unsolved obstacles for most physicians in treating the lung cancer patients. The overall treatment outcome of lung is still poor. About 30% of patients with NSCLC present early stage disease and receive potentially curative treatment, but up to 40% of these patients relapse within 5 years (Hoffman et al. 2000, Mountain 1997, Naruke et al. 1988). Recent genetic epidemiology and pharmacogenomic studies revealed that lung cancer in different ethnic group, particularly in East Asia, female non-smoker lung adenocarcinoma, is a distinct disease and may be different from patients from western country (Yang et al. 2008). The current clinicopathological staging system for NSCLC is inadequate for predicting whether these patients, with the same clinical and pathological features, will develop metastasis or would be cured. To improve the management of NSCLC, molecular methods for identifying patients at high risk of recurrence of metastasis after resection are needed.

Key words. Non-small cell lung cancer, biomarker, predictive, prognostic, microRNA

The biomarker can be classified into three types. The prognostic markers are biomarkers that can be used to estimate patients' outcome independent of therapeutic decision. The predictive markers are biomarkers that can assist therapeutic decision for the patients. The diagnostic biomarkers are biomarkers that can assist disease diagnosis, disease subclassification and monitoring of the therapeutic response for the patients.

Biomarkers regarded as diagnostic and prognostic factors for cancer has been presented in many studies (Ludwig and Weinstein 2005, Sidransky 2002). Unfortunately, there is no mature biomarker used in clinical practice. For example, the concentration of prostate-specific antigen (PSA) is higher in patient with prostate cancer than in normal one. However, the false positive rate is high when using PSA as a diagnostic factor for prostate cancer. This phenomenon may be due to benign prostate hyperplasia and inflammation *etc*. Until today, it is still a controversial issue of using PSA in prostate cancer screening (Etzioni et al. 2002, Vastag 2002).

The processes of normal cell transforming into cancer cell are complex. There are three major levels in these phenomena: genomic DNA, DNA transcribes to RNA, and RNA translates to protein (Ludwig and Weinstein 2005, Sidransky 2002, Vastag 2002, Esquela-Kerscher and Slack 2006). We measure the changes of elements in the above three levels as biomarkers. For examples, tumorigenesis may due to the increase of the oncogene copy number and the deletion of the tumor suppress gene in DNA level, differential gene expressions of RNA can be found between normal and cancer cells; and abnormal expressions or structures of proteins could be detected in cancer cells.

Recently, the definitions of classical oncogenes and tumor suppressors have been expanded to include a new species of RNAs, known as microRNAs, due to the epochal findings on regulatory functions of non-protein-coding RNAs. The microRNAs are a new class of small non-protein-coding RNAs that can act as endogenous RNA interference (Hammond 2006). The microRNAs can post-transcriptionally regulate hundreds of their target genes expressions and control a wide range of biological functions such as cellular proliferation, differentiation and apoptosis (Esquela-Kerscher and Slack 2006). Recent evidence indicates that microRNAs may function as tumor suppressors or oncogenes; and alterations in microRNA expression may play a critical role in the cancer initiation and progression (Calin and Croce 2006).

The microRNAs are single-stranded RNAs consist of 18-25 nucleotides, and could be found in both plant and animal cells. They regulate their targets by direct-cleavage of the mRNA or by inhibition of protein synthesis, according to the degree of complementarities with their targets' 3'-untranslated regions (3'-UTR)'. Perfect or nearly perfect base pairings between microRNA and its target 3'-UTR induce RNA-induced silencing complex RISC to cleave the target mRNA, whereas imperfect base pairing induces mainly translational silencing of the target but can also reduce the amount of the transcript of target (He and Hannon 2004).

Microarray technology (Hoheisel 2006, Schena et al. 1995, Schena et al. 1996) is commonly used to measure the mRNA expressions in large-scale. Recently, microarray gene expression profiles were shown to be correlated with cancer behavior or clinical outcome and to provide a better prediction of patient survival than traditional methods (Ludwig and Weinstein 2005, Golub et al. 1999, Ramaswamy et al. 2003), including lung cancer (Beer et al. 2002, Bhattacharjee et al. 2001, Garber et al. 2001, Gordon et al. 2002, Wigle et al. 2002). However, the use of microarrays in clinical practice is limited by technical matters: the large number of genes employed in gene profiling (Ramaswamy 2004), complicated methods, and the lack of reproducibility and independent validation. The genes selected for profiling in different studies of lung cancer have varied considerably, and only a few genes have been consistently included in the analyses (Beer et al. 2002, Bhattacharjee et al. 2001, Garber et al. 2001, Wigle et al. 2002) . Gene expression profiles can vary according to the microarray platform and analytical strategy used in the analysis (Lossos et al. 2004) .

The microRNA expression profiles may be useful for the classification, diagnosis or prognosis of some human malignancies (Esquela-Kerscher and Slack 2006, Calin and Croce 2006). Furthermore, the recent reports revealed that the expression profiling of microRNAs may be a more accurate method of classifying cancer subtype than using the expression profiles of protein coding genes (Esquela-Kerscher and Slack 2006, Volinia et al. 2006) . So far, our understanding of microRNA expression patterns in cancer patients is just starting to emerge. However, several microRNAs have been reported to be associated with the clinical outcome, including chronic lymphocytic leukemia (Calin et al. 2005), breast (Iorio et al. 2005), pancreas (Roldo et al. 2006) and lung (Takamizawa et al. 2004, Yanaihara et al. 2006, Yu et al. 2008) cancers.

After microarray analysis, the gene expression profiles usually require further validation, preferably by reverse-transcription polymerase chain reaction (RT-PCR). The RT-PCR method using paraffin-embedded pathology specimen is reproducible and is applicable in clinical practice. However, RT-PCR can only be used to screen for a small number of genes (Ramaswamy 2004) . A recent study demonstrated good prediction of prognosis by real-time RT-PCR lung cancer (Chen et al. 2007, Endoh et al. 2004); still, whether real-time RT-PCR gene signatures can predict survival of all NSCLC patients is uncertain.

The current studies showed similar results with the previous reports in mRNA expression (Beer et al. 2002, Bhattacharjee et al. 2001, Garber et al.

2001, Wigle et al. 2002, Lossos et al. 2004, Endoh et al. 2004), or in microRNA expression (Calin and Croce 2006, Calin et al. 2005, Yanaihara et al. 2006) . Nevertheless, the limitations of current studies exist, including too many genes (more than 10 genes), lack of RT-PCR confirmation, lack of external validation, and single gene use only.

In the studies of biomarker development, too many genes used in a predictive model may cause difficulty in clinical application; an example is Microarray technology: the expression profiles consist of many genes. In addition, the results obtained from microarray experiments are not completely reliable and further confirmations by other approaches are required, such as RT-PCR or Northern blot. The internal validation approach such as cross-validation or training-testing approach for model validation is not enough. The external validation which contains the sample not overlapping the population of which the model derived from is necessary (Simon et al. 2003). If external validation approach is not performed, the predictive model may not be applied to new samples. Using single gene to predict the risk of patients is a conventional and yet controversial method. The current studies reported that using single gene to classify patients must collocate with strong association with outcome, for example, odds ratio more than 200. The sum of weighted contribution of genes, such as risk score method, can resolve this problem (Wang et al. 2006, Ware 2006) . The predictive models by five genes (Chen et al. 2007) or five microRNAs (Yu et al. 2008) are based on the sum of weighted contribution of each gene or microRNA, and explained complex disease by different pathways. In addition, the models are further verified in the external cohorts. The findings of this dissertation showed potential biomarkers of prognosis of lung cancer. If patients are predicted as high risk, they will need more clinical treatments even they are in early stage of the disease. These biomarkers can also be used to drug development. Inhibited expression of risk genes or enhanced expression of protective genes may lead the patients from high risk to low risk. However, more evidences are needed for these biomarkers obtained from current studies to ensure that they are useful. The large prospective trial, including multiple medical centers, is needed.

References

Beer DG, Kardia SL, Huang CC, Giordano TJ, Levin AM, Misek DE, Lin L, Chen G, Gharib TG, Thomas DG, Lizyness ML, Kuick R, Hayasaka S, Taylor JM, Iannettoni MD, Orringer MB,Hanash S (2002) Gene-expression profiles predict survival of patients with lung adenocarcinoma. Nat Med 8: 816-824

Bhattacharjee A, Richards WG, Staunton J, Li C, Monti S, Vasa P, Ladd C,

Beheshti J, Bueno R, Gillette M, Loda M, Weber G, Mark EJ, Lander ES, Wong W, Johnson BE, Golub TR, Sugarbaker DJ,Meyerson M (2001) Classification of human lung carcinomas by mRNA expression profiling reveals distinct adenocarcinoma subclasses. Proc Natl Acad Sci U S A 98: 13790-13795

Calin GA, Ferracin M, Cimmino A, Di Leva G, Shimizu M, Wojcik SE, Iorio MV, Visone R, Sever NI, Fabbri M, Iuliano R, Palumbo T, Pichiorri F, Roldo C, Garzon R, Sevignani C, Rassenti L, Alder H, Volinia S, Liu CG, Kipps TJ, Negrini M,Croce CM (2005) A MicroRNA signature associated with prognosis and progression in chronic lymphocytic leukemia. N Engl J Med 353: 1793-1801

Calin GA,Croce CM (2006) MicroRNA-cancer connection: the beginning of a new tale. Cancer Res 66: 7390-7394

Chen HY, Yu SL, Chen CH, Chang GC, Chen CY, Yuan A, Cheng CL, Wang CH, Terng HJ, Kao SF, Chan WK, Li HN, Liu CC, Singh S, Chen WJ, Chen JJ,Yang PC (2007) A five-gene signature and clinical outcome in non-small-cell lung cancer. N Engl J Med 356: 11-20

Endoh H, Tomida S, Yatabe Y, Konishi H, Osada H, Tajima K, Kuwano H, Takahashi T,Mitsudomi T (2004) Prognostic model of pulmonary adenocarcinoma by expression profiling of eight genes as determined by quantitative real-time reverse transcriptase polymerase chain reaction. J Clin Oncol 22: 811-819

Esquela-Kerscher A,Slack FJ (2006) Oncomirs - microRNAs with a role in cancer. Nat Rev Cancer 6: 259-269

Etzioni R, Penson DF, Legler JM, di Tommaso D, Boer R, Gann PH,Feuer EJ (2002) Overdiagnosis due to prostate-specific antigen screening: lessons from U.S. prostate cancer incidence trends. J Natl Cancer Inst 94: 981-990

Garber ME, Troyanskaya OG, Schluens K, Petersen S, Thaesler Z, Pacyna-Gengelbach M, van de Rijn M, Rosen GD, Perou CM, Whyte RI, Altman RB, Brown PO, Botstein D,Petersen I (2001) Diversity of gene expression in adenocarcinoma of the lung. Proc Natl Acad Sci U S A 98: 13784-13789

Golub TR, Slonim DK, Tamayo P, Huard C, Gaasenbeek M, Mesirov JP, Coller H, Loh ML, Downing JR, Caligiuri MA, Bloomfield CD,Lander ES (1999) Molecular classification of cancer: class discovery and class prediction by gene expression monitoring. Science 286: 531-537

Gordon GJ, Jensen RV, Hsiao LL, Gullans SR, Blumenstock JE, Ramaswamy S, Richards WG, Sugarbaker DJ,Bueno R (2002) Translation of microarray data into clinically relevant cancer diagnostic tests using gene expression ratios in lung cancer and mesothelioma. Cancer Res 62: 4963-4967

Hammond SM (2006) MicroRNAs as oncogenes. Curr Opin Genet Dev 16: 4-9

He L,Hannon GJ (2004) MicroRNAs: small RNAs with a big role in gene regulation. Nat Rev Genet 5: 522-531

Hoffman PC, Mauer AM,Vokes EE (2000) Lung cancer. Lancet 355: 479-485

Hoheisel JD (2006) Microarray technology: beyond transcript profiling and genotype analysis. Nat Rev Genet 7: 200-210

Iorio MV, Ferracin M, Liu CG, Veronese A, Spizzo R, Sabbioni S, Magri E, Pedriali M, Fabbri M, Campiglio M, Menard S, Palazzo JP, Rosenberg A, Musiani P, Volinia S, Nenci I, Calin GA, Querzoli P, Negrini M,Croce CM (2005) MicroRNA gene expression deregulation in human breast cancer. Cancer Res 65: 7065-7070

Jemal A, Siegel R, Ward E, Murray T, Xu J, Smigal C,Thun MJ (2006) Cancer statistics, 2006. CA Cancer J Clin 56: 106-130

Lossos IS, Czerwinski DK, Alizadeh AA, Wechser MA, Tibshirani R, Botstein D,Levy R (2004) Prediction of survival in diffuse large-B-cell lymphoma based on the expression of six genes. N Engl J Med 350: 1828-1837

Ludwig JA,Weinstein JN (2005) Biomarkers in cancer staging, prognosis and treatment selection. Nat Rev Cancer 5: 845-856

Mountain CF (1997) Revisions in the International System for Staging Lung Cancer. Chest 111: 1710-1717

Naruke T, Goya T, Tsuchiya R,Suemasu K (1988) Prognosis and survival in resected lung carcinoma based on the new international staging system. J Thorac Cardiovasc Surg 96: 440-447

Pepe MS, Janes H, Longton G, Leisenring W,Newcomb P (2004) Limitations of the odds ratio in gauging the performance of a diagnostic, prognostic, or screening marker. Am J Epidemiol 159: 882-890

Ramaswamy S (2004) Translating cancer genomics into clinical oncology. N Engl J Med 350: 1814-1816

Ramaswamy S, Ross KN, Lander ES,Golub TR (2003) A molecular signature of metastasis in primary solid tumors. Nat Genet 33: 49-54

Roldo C, Missiaglia E, Hagan JP, Falconi M, Capelli P, Bersani S, Calin GA, Volinia S, Liu CG, Scarpa A,Croce CM (2006) MicroRNA expression abnormalities in pancreatic endocrine and acinar tumors are associated with distinctive pathologic features and clinical behavior. J Clin Oncol 24: 4677-4684

Schena M, Shalon D, Davis RW,Brown PO (1995) Quantitative monitoring of gene expression patterns with a complementary DNA microarray. Science 270: 467-470

Schena M, Shalon D, Heller R, Chai A, Brown PO,Davis RW (1996) Parallel human genome analysis: microarray-based expression monitoring of 1000 genes. Proc Natl Acad Sci U S A 93: 10614-10619

Sidransky D (2002) Emerging molecular markers of cancer. Nat Rev Cancer 2: 210-219

Simon R, Radmacher MD, Dobbin K,McShane LM (2003) Pitfalls in the use of DNA microarray data for diagnostic and prognostic classification. J Natl Cancer Inst 95: 14-18

Takamizawa J, Konishi H, Yanagisawa K, Tomida S, Osada H, Endoh H, Harano T, Yatabe Y, Nagino M, Nimura Y, Mitsudomi T,Takahashi T (2004) Reduced expression of the let-7 microRNAs in human lung cancers in association with shortened postoperative survival. Cancer Res 64: 3753-3756

Vastag B (2002) Study concludes that moderate PSA levels are unrelated to prostate cancer outcomes. Jama 287: 969-970

Volinia S, Calin GA, Liu CG, Ambs S, Cimmino A, Petrocca F, Visone R, Iorio M, Roldo C, Ferracin M, Prueitt RL, Yanaihara N, Lanza G, Scarpa A, Vecchione A, Negrini M, Harris CC,Croce CM (2006) A microRNA expression signature of human solid tumors defines cancer gene targets. Proc Natl Acad Sci U S A 103: 2257-2261

Wang TJ, Gona P, Larson MG, Tofler GH, Levy D, Newton-Cheh C, Jacques PF, Rifai N, Selhub J, Robins SJ, Benjamin EJ, D'Agostino RB,Vasan RS (2006) Multiple biomarkers for the prediction of first major cardiovascular events and death. N Engl J Med 355: 2631-2639

Ware JH (2006) The limitations of risk factors as prognostic tools. N Engl J Med 355: 2615-2617

Wigle DA, Jurisica I, Radulovich N, Pintilie M, Rossant J, Liu N, Lu C, Woodgett J, Seiden I, Johnston M, Keshavjee S, Darling G, Winton T, Breitkreutz BJ, Jorgenson P, Tyers M, Shepherd FA,Tsao MS (2002) Molecular profiling of non-small cell lung cancer and correlation with disease-free survival. Cancer Res 62: 3005-3008

Yanaihara N, Caplen N, Bowman E, Seike M, Kumamoto K, Yi M, Stephens RM, Okamoto A, Yokota J, Tanaka T, Calin GA, Liu CG, Croce CM,Harris CC (2006) Unique microRNA molecular profiles in lung cancer diagnosis and prognosis. Cancer Cell 9: 189-198

Yang CH, Yu CJ, Shih JY, Chang YC, Hu FC, Tsai MC, Chen KY, Lin ZZ, Huang CJ, Shun CT, Huang CL, Bean J, Cheng AL, Pao W, Yang PC (2008) Specific EGFR mutations predict treatment outcome of stage IIIB/IV patients with chemotherapy-naive non-small-cell lung cancer receiving first-line gefitinib monotherapy. J Clin Oncol 26:2745-53

Yu SL, Chen HY, Chang GC, Chen CY, Chen HW, Singh S, Cheng CL, Yu CJ, Lee YC, Chen HS, Su TJ, Chiang CC, Li HN, Hong QS, Su HY, Chen CC, Chen WJ, Liu CC, Chan WK, Chen WJ, Li KC, Chen JJ,Yang PC (2008) MicroRNA signature predicts survival and relapse in lung cancer. Cancer Cell 13: 48-57

Screening of Biomarker for Oral Squamous Cell Carcinoma Correlating with Clinicopathological Findings Using by Laser Microdissection and cDNA Microarray

Tarou Irié, Gou Yamamoto, Tomohide Isobe, Taku Matsunaga and Tetsuhiko Tachikawa

Department of Oral Pathology, Showa University School of Dentistry
1-5-8 Hatanodai, Shinagawa-ku, Tokyo, 142-8555 Japan.

Summary. Infiltrating pattern is one of important histopathological prognostic factor in head and neck squamous carcinomas. In this study we have assessed differentially expressed genes in each infiltrating patterns by laser microdissection and cDNA microarray. We classified infiltrating patterns of oral squamous cell carcinoma into major three groups; INFα, INFβ, and INFγ, based on Japanese General Rule for Clinical and Pathological Studies on Cancer, and collected cancer cells of each infiltrating pattern by laser microdissection. Total RNA was extracted from collected cells and was reverse-transcribed. PCR-based amplification of cDNA, cDNA probe labeling and array hybridization were conducted. We compared expression profiles of cancer cells of each infiltrating pattern to that of normal mucosal epithelium using cDNA microarray consisting of 8327 human genes. We identified several genes that were commonly up- or downregulated in each infiltrating pattern. The differentially expressed genes were growth-related proteins, oncogenes, enzymes and nuclear proteins. The list of genes identified in these analyses provides potentially valuable information for preoperative assessment of evaluation of malignancy and represents a source of novel targets for cancer therapy.

Key words. laser microdissection, cDNA microarray, oral squamous cell carcinoma, gene expression profile, metastasis

Introduction

Many clinical and histopathological factors have prognostic significance in patients with oral squamous cell carcinoma. In 1986 the Departments of Pathology at the Karolinska Institute and the M. D. Anderson Hospital proposed the system which assessed six histomorphologic tumor features.

These variables were degree of keratinization, nuclear pleomorphism, number of mitoses, pattern of invasion, stage of invasion, and leukocyte infiltration. Recently, invasive front areas of squamous cell carcinoma are more important for its invasive and metastatic capacity and thus include more prognostic information than central and superficial parts of the same tumor (Bryne 1998). Many other authors have reported that a significant correlation between pattern of invasion and distant metastasis, as well as overall survival (Brandwein-Gensler et al. 2005, Spiro et al. 1999) and also local recurrence (Brandwein-Gensler et al. 2005). It is still unclear what molecules participate in the formation of various pattern of invasion of oral squamous cell carcinoma. In Japan, Japanese General Rule for Clinical and Pathological Studies on Cancer are routinely used by pathologists on making pathological diagnostic reports for various cancers. Based on this rule, we classified into three major groups of infiltrating pattern at invasive front area of oral squamous cell carcinoma; INFα, INFβ, and INFγ and examined genes correlating with infiltrating patterns by means of laser microdissection and cDNA microarray.

Materials and methods

Tissue samples and specimen selection

Specimens of primary oral squamous cell carcinoma (OSCC) were used for the present analysis. Histological examination of infiltrating pattern was performed according to Japanese General Rule for Clinical and Pathological Studies on Cancer (for esophagus, stomach and colon, etc). Histological criteria are as follows: INFα; cancer shows expansive growth with clear boundary between surrounding tissue. INFβ; infiltration pattern shows the condition between α and β. INFγ; cancer shows infiltrative growth with unclear boundary between surrounding tissue.

Preparation of frozen fixed samples and laser microdissection

Tissue sampled from resected materials were embedded in OCT compound and frozen in isopentane cooled in liquid nitrogen. The frozen blocks were sliced by a cryomicrotome at a thickness of 8 μm, and each tissue section was affixed to a slide to which an original thin film (provided by Meiwa Shoji Co, LTD, Tokyo, Japan) had been attached by silicone adhesive (GE Toshiba Silicone, Tokyo, Japan).

Slice samples were stored at –40°C until use. The sliced sample was quickly fixed in 100% methanol for 3 minutes, and then returned to room temperature and stained with 1% toluidine blue solution (Irié et al. 2004).

Laser Microbeam System (P.A.L.M, Bernried, Germany) using 337 nm nitrogen laser was used for laser microdissection. We procured a few hundred cells from cancer tissues of each infiltrating pattern and normal mucosal epithelia in each case. We examined three samples of gene expression in each of infiltrating pattern.

RNA extraction from microdissected samples

Total RNA was independently extracted from each population of laser-microdissected cells. Briefly, the microdissected cells within the cap were covered with 200 μl buffer solution, 4 M guanidine thiocyanate, 25 mM sodium citrate, and 0.5% sarcosyl, and the cap was placed on the tube and vortexed. After the addition of 20 μl of 2 M sodium acetate, 220 μl of water-saturated phenol and 60 μl of chloroform-isoamyl alcohol, the tube was centrifuged at 10000 g at 4°C for 30 minutes to separate the aqueous and organic phases. The aqueous layer was transferred to a new tube. Two μl of glycogen and 200 μl of isopropanol were added and centrifuged at 10000 g at 4°C for 30 minutes. After removing the majority of the supernatant, the pellet was washed with 70% ethanol. After the pellet was centrifuged and air-dried, the mRNA was resuspended in RNase-free water.

cDNA synthesis and amplification

Total RNA was reverse-transcribed using the SMART PCR cDNA Synthesis Kit (Clontech, CA, USA). Four μl of total RNA, 1μl of cDNA Synthesis (CDS) Primer, and 1 μl of SMART II oligonucleotide were mixed and incubated at 70°C for 8 minutes. After short spinning, 2 minutes on ice and 2 minutes at 42°C, a master mix containing 2 μl of 5× buffer, 1 μl dithiothreitol (20 mmol/L), 1 μl dNTP (10 mmol/L) and 0.5 μl RNase H^- Moloney murine leukemia virus reverse transcriptase (PowerScript, Clontech, CA, USA) were added and incubated at 42°C for 1 hour. Afterward, cDNA was mixed with 38.5 μl of TE buffer and purified by the Atlas NucleoSpin Extraction Kits (Clontech, CA, USA). Therefore, 400 μl of buffer NT2 were added to the cDNA to load a column. According to the manufacture's protocol, the column was washed three times. For elution, 50 μl of elution buffer were applied, incubated for 2 minutes, and centrifuged.

From the eluted cDNA, 2 μl were separated for further determination of 260 nm absorbance. For the PCR-based amplification, the remaining 42 μl of cDNA were mixed with 5 μl of 10× buffer, 1 μl PCR primer (10 μmol/L), 1 μl dNTP (10 mmol/L), and 1 μl Advantage 2 polymerase mix.

PCR conditions were 95°C for 1 minute, followed by 22 cycles with 95°C for 15 seconds, 65°C for 30 seconds, and 68°C for 3 minutes. The resulting PCR product was purified using the Atlas NucleoSpin Extraction Kits as described above. All incubations were performed with GeneAmp 2400 PCR cycler (PE Applied Biosystems, CA, USA).

Probe labeling and Array hybridization

For array hybridization, we used nylon filters with 1176 spotted cDNA (Human cancer 1.2 Atlas cDNA array, Clontech, CA, USA). The purified PCR product was labeled with α-^{32}P dATP using the Atlas SMART Probe Amplification Kit (Clontech, CA, USA). Thirty-two μl of PCR product and 1 μl of CDS primer were heated at 95°C for 8 minutes. After 2 minutes at 50°C, a master mix containing 5 μl of 10× labeling buffer, 5 μl of dNTP (without dATP), 5 μl of α-^{32}P dATP (Amersham Pharmacia Biotech, Tokyo, Japan), and 1 μl of Klenow enzyme were mixed, and incubated for 30 minutes at 50°C. Reaction was stopped by applying 2 μl of 0.5-mol/L ethylenediaminetetraacetic acid. Labeled DNA was purified by Atlas NucleoSpin Extraction Kits as described above, eluted with 100 μl of elution buffer, and resulted in ~5 to 8 × 10^6 cpm. Afterward, array hybridization was performed according to the protocol. Filters with ^{32}P labeled PCR product were incubated at 68°C overnight. They were washed three times in 200 ml of 2× standard saline citrate and 1% sodium dodecyl sulfate at 68°C for 30 minutes. Finally, they were wrapped in plastic and exposed to an imaging plate (Fuji Photo Film, Tokyo, Japan) in lead sheathing. Imaging plate was read with a phosphorimaging system (STORM 830, Molecular Dynamic Inc, Tokyo, Japan).

Array analysis

Analysis was performed using the AtlasImage 2.01 software (Clontech, CA, USA). Global normalization was calculated by the sum method. For both arrays, differences of signal intensity minus background were added for all values over background. Afterward, normalization coefficient was determined by calculating the ratio of array 1 sum and array 2 sum. After normalization, background was subtracted, ratio threshold was set at 4 (Irié et al. 2004).

Results

We collected cancer cells by laser microdissection and analyzed three samples each about each infiltrating pattern. Genes that were commonly up or down-regulated (more than 4-fold up regulated or less than 0.25-fold) in all samples of each infiltrating pattern compared to normal mucosal epithelium were examined. Three genes were commonly up-regulated and 3 were commonly down-regulated in INFα (Table 1).

Table 1 Commonly up or downregulated genes in INFα

Gene name	Classification
Up-regulated genes	
activator 1 37-kDa subunit; replication factor C 37-kDa subunit (RFC37)	DNA replication
interferon gamma-induced protein precursor (gamma-IP10)	growth factor
CXC chemokine precursor (CXCL13)	growth factor
Down-regulated genes	
CDC-like kinase 3 (CLK3)	cell cycle-regulating kinase
interleukin-1 receptor antagonist protein (IL-1Ra)	growth factor
neutrophil gelatinase-associated lipocalin (NGAL)	oncogene

In INFβ, commonly up-regulated genes were five and commonly down-regulated genes were 3 (Table 2).

Table 2 Commonly up or downregulated genes in INFβ

Gene name	Classification
Up-regulated genes	
vascular endothelial growth factor receptor 1 (VEGFR1)	growth factor
c-jun N-terminal kinase 2 (JNK2)	intercellular kinase
fuse-binding protein 3 (FBP3)	transcription factor
replication factor C 38-kDa subunit (RFC38)	DNA polymerase
interferon gamma-induced protein precursor (gamma-IP10)	growth factor
Down-regulated genes	
neutrophil elastase inhibitor	protease inhibitor
cytokeratin 10	cytoskeltal protein
neutrophil gelatinase-associated lipocalin (NGAL)	oncogene

Three were commonly up-regulated and 2 were commonly down-regulated in INFγ (Table 3). Interferon gamma-induced protein precursor (gamma-IP10) was detected as commonly up-regulated gene in both INFα and INFβ. Vascular endothelial growth factor receptor 1 (VEGFR1) was detected as commonly up-regulated gene in both INFβ and INFγ. Neutrophil gelatinase-associated lipocalin precursor (NGAL) was detected as commonly down-regulated gene in all infiltrating patterns.

Table 3 Commonly up or downregulated genes in INFγ

Gene name	Classification
Up-regulated genes	
vascular endothelial growth factor receptor 1 (VEGFR1)	growth factor
rhoC (H9); small GTPase (rhoC)	oncogene
centromere protein F (CENP-F)	chromatin protein
Down-regulated genes	
cyclin-dependent kinase inhibitor 1 (CDKN1A, p21, Cip1)	CDK inhibitor
neutrophil gelatinase-associated lipocalin (NGAL)	oncogene

Discussion

The array identified a total of 19 genes that included growth-related proteins, oncogenes, enzymes and nuclear proteins. Replication factor C is a multi-protein complex consisting of five different polypeptides. It recognizes a primer on a template DNA, binds to a primer terminus, and helps load proliferating cell nuclear antigen onto the DNA template. Replication factor C 37-kDa subunit (RFC37, RFC2) is third-largest subunit of the RF-C complex and is required not only for chromosomal DNA replication but also for a cell cycle checkpoint (Noskov et al. 1998). The gains of RFC2 (7q11) that were detected by comparative genomic hybridization in glioblastomas (Nakahara et al. 2004) and renal cell carcinoma cell lines (Speicher et al. 1994). Interferon gamma-induced protein precursor (gamma-IP10, INP10, CXCL10) may be an important mediator of the inflammatory response to interferons. Angiolillo et al. reported that IP10 may participate in the regulation of angiogenesis during inflammation and tumorigenesis (Angiolillo et al. 1995). Zhang et al. demonstrated that gamma-IP10 is a RAS target gene and is overexpressed in the majority of colorectal cancers (Zhang et al. 1997). Over-expression of INP10 was revealed in malignant pleural mesotheliomas (Kettunen et al. 2005). Moreover, gamma-IP10 has the inhibitory effects on the experimental tumor metastases of a mammary carcinoma cell line 4T1 in vivo (Yang et al. 2006). It may be responsible for better prognosis of INFα than other infiltrating pattern. CXC chemokine precursor is identified a protein, termed B cell–attracting chemokine 1 (BCA-1, CXCL13), that is chemotactic for human B lymphocytes (Legler et al. 1998). This mature protein shares 23–34% identical amino acids with known CXC chemokines and is constitutively expressed in secondary lymphoid organs. It is demonstrated that CXCL13 is correlated with lymphoproliferative disorders (Mori et al. 2003) and malignant lymphoma (Dupuis et al. 2006). Hofman et al. (2006) reported that up-regulation of CXCL13 gene in gastric cancer tissue associated with Helicobacter pylori infection. IL-1 receptor antagonist (IL-1Ra) is the naturally occurring inhibitor of IL-1. Dionne S et al. (1998) examined that the balance between IL-1 and IL-1Ra in inflammatory bowel disease (IBD) by measuring their secretion by cultured explants of colonic biopsies and demonstrated a decreased IL-1Ra/IL-1 ratio in involved IBD tissue. Down-regulation of IL-1Ra in INFα causes inflammatory response to cancer cells and may participate in better prognosis of INFα. Neutrophil gelatinase-associated lipocalin (NGAL), also known as lipocalin 2, is a 25-kDa lipocalin initially purified from neutrophil granules. It is thought to play a role in regulating cellular growth since its expression is highly upregulated in a variety of human cancers, and lipocalin ligands have been shown to regulate proliferation, differentiation, and protease activities (Bratt 2000). However, in striking contrast, a recent study by Hanai et al. (2005) showed that introduction of NGAL into Ras-transformed 4T1 mouse mammary tumor cells enhanced the epithelial phenotype, reduced tumor growth, and suppressed metastasis. It is thought that the actions of NGAL may be heterogeneous and specific to the cancer cell type (Devarajan 2007). In oral mucosal epithelium, NGAL may serve as protective role.

As to INFβ, vascular endothelial growth factor (VEGF) and its receptors fms-like tyrosine kinase-1 (Flt-1, VEGFR1) play critical roles in angiogenesis and tumor progression. VEGF and VEGFR1 are widely expressed in nasopharyngeal carcinoma (NPC) tissues, and are positively related to clinical features and prognosis of NPC patients (Sha et al. 2006). Lalla et al. (2003) demonstrated that VEGFR-1 was significantly higher in tumor cells and macrophages than in vascular endothelial cells and speculated that VEGF may be an autocrine regulator of tumor cell activity in addition to its known angiogenic effects on vascular endothelial cells. c-Jun NH2-terminal kinases (JNK) are members of the mitogen-activated protein kinase family that, when stimulated, regulate a variety of cellular activities, including proliferation, differentiation, and tumorigenesis. There are three JNK genes, JNK1 and JNK2 gene expression is ubiquitous and JNK3 is restricted to the heart, testis, and brain. Gross et al. (2007) clarified that JNK activity was increased in human HNSCC compared with normal-appearing epithelium. Replication factor C 38-kDa subunit (RFC38) is synonym for RFC5. RFC5 is up-regulated in HPV-positive squamous cell carcinoma of the head and neck. Neutrophil elastase inhibitor suppresses the proliferation, motility and chemotaxis of a pancreatic carcinoma cell line (Kamohara et al. 1997) and have ability to inhibit spontaneous pulmonary metastasis of human lung carcinoma cell line (Inada et al. 1998). Robinson et al. (1996) demonstrated that expression of neutrophil elastase inhibitor was increased in keratinizing epithelia of normal oral mucosa, suprabasal layers of dysplastic oral epithelia and in well-differentiated squamous cell carcinoma but not in non-keratinizing normal mucosa and basal cell carcinoma. These reports may reflect our results that indicated down-regulation of neutrophil elastase inhibitor, because carcinoma cells of INFβ show decreasing of keratinization compared to that of INFα. Hansson at al. (2001) examine keratin expression in organotypic epithelia with normal, immortalized and malignant human buccal cells. They showed K4/K13, K1/K10, K6/K16 were variably expressed in normal and immortalized buccal cells but were not detected in malignant one. Carrilho et al. (2004) observed that expression of keratins 8 and 17 and loss of keratins 10 and 13 are good markers of malignant transformation in human cervical mucosa.

As for INFγ, Rho family members are known to regulate malignant transformation and motility of cancer cells. Liu et al. (2007) showed that RhoC was positively expressed in metastatic cancer tissues even higher than that in primary gastric cancer tissues, and overexpression of RhoC GTPase in GES-1 cells could produce the motile and invasive phenotype. O'Brien et al. (2007) found that CENP-F up-regulation was significantly associated with worse overall survival and reduced metastasis-free survival of breast cancer patients. Cyclin-dependent kinase inhibitor 1 (CDKN1A, p21, Cip1) is a recently identified gene involved in cell cycle regulation through cyclin-CDK-complex inhibition. CDKN1A expression was strongly associated with squamous cell differentiation of carcinomas, because undifferentiated carcinomas showed very low levels of CDKN1A expression, whereas carcinomas with squamous cell differentiation had normal or high levels of CDKN1A expression (Nadal et al. 1997). This would be compatible with our results, because cancer cells of INFγ show always poorly differentiation.

This is the first report that examined gene expression correlating infiltrating pattern of oral squamous cell carcinoma by means of laser microdissection and cDNA microarray. The list of genes identified in these analyses provides potentially valuable information for preoperative assessment of evaluation of malignancy and represents a source of novel targets for cancer therapy.

References

Angiolillo AL, Sgadari C, Taub DD, Liao F, Farber JM, Maheshwari S, Kleinman HK, Reaman GH, Tosato G (1995) Human interferon-inducible protein 10 is a potent inhibitor of angiogenesis in vivo. J Exp Med 182:155-162

Brandwein-Gensler M, Teixeira MS, Lewis CM, Lee B, Rolnitzky L, Hille JJ, Genden E, Urken ML, Wang BY (2005) Oral squamous cell carcinoma. Histologic risk assessment, but not margin status, is strongly predictive of local disease-free and overall survival. Am J Surg Pathol 29:167–178

Bratt T (2000) Lipocalins and cancer. Biochim Biophys Acta 1482:318–326

Bryne M (1998) Is the invasive front of an oral carcinoma the most important area for prognostication? Oral Dis 4:70 –77

Carrilho C, Alberto M, Buane L, David L (2004) Keratins 8, 10, 13, and 17 are useful markers in the diagnosis of human cervix carcinomas. Hum Pathol 35:546-551

Devarajan P (2007) Neutrophil gelatinase-associated lipocalin: new paths for an old shuttle. Cancer Ther 5:463-470

Dionne S, D'Agata ID, Hiscott J, Vanounou T, Seidman EG (1998) Colonic explant production of IL-1 and its receptor antagonist is imbalanced in inflammatory bowel disease (IBD). Clin Exp Immunol 112: 435–442

Dupuis J, Boye K, Martin N, Copie-Bergman C, Plonquet A, Fabiani B, Baglin AC, Haioun C, Delfau-Larue MH, Gaulard P (2006) Expression of CXCL13 by neoplastic cells in angioimmunoblastic T-cell lymphoma (AITL): a new diagnostic marker providing evidence that AITL derives from follicular helper T cells. Am J Surg Pathol 30:490-494

Folkesson M, Kazi M, Zhu C, Silveira A, Hemdahl AL, Hamsten A, Hedin U, Swedenborg J, Eriksson P (2007) Presence of NGAL/MMP-9 complexes in human abdominal aortic aneurysms. Thromb Haemost 98:427-433

Gross ND, Boyle JO, Du B, Kekatpure VD, Lantowski A, Thaler HT, Weksler BB, Subbaramaiah K, Dannenberg AJ (2007) Inhibition of Jun NH2-terminal kinases suppresses the growth of experimental head and neck squamous cell carcinoma. Clin Cancer Res 13:5910-5917

Hanai J, Mammoto T, Seth P, Mori K, Karumanchi A, Barasch J, Sukhatme VP (2005) Lipocalin 2 diminishes invasiveness and metastasis of ras-transformed cells. J Biol Chem 280:13641–13647

Hansson A, Bloor BK, Haig Y, Morgan PR, Ekstrand J, Grafström RC (2001) Expression of keratins in normal, immortalized and malignant oral epithelia in organotypic culture. Oral Oncol 37:419-430

Hofman VJ, Moreilhon C, Brest PD, Lassalle S, Le Brigand K, Sicard D, Raymond J, Lamarque D, Hébuterne XA, Mari B, Barbry PJ, Hofman PM (2007) Gene expression profiling in human gastric mucosa infected with Helicobacter pylori. Mod Pathol 20:974-89

Inada M, Yamashita J, Nakano S, Ogawa M (1998) Complete inhibition of spontaneous pulmonary metastasis of human lung carcinoma cell line EBC-1 by a neutrophil elastase inhibitor (ONO-5046.Na). Anticancer Res 18:885-890

Irié T, Aida T, Aida M, Nagoshi Y, Tsuchiya R, Yamamoto G, Maeda Y, Saito M, Tachikawa T (2004) Laser pressure cell transfer method: A new microdissection technique for frozen sections. Oral Med Pathol 9:53-60

Irié T, Aida T, Tachikawa T (2004) Gene expression profiling of oral squamous cell carcinoma using laser microdissection and cDNA microarray. Med Electron Microsc 37:89-96

Kamohara H, Sakamoto K, Mita S, An XY, Ogawa M (1997) Neutrophil elastase inhibitor (ONO-5046.Na) suppresses the proliferation, motility and chemotaxis of a pancreatic carcinoma cell line, Capan-1. Res Commun Mol Pathol Pharmacol 98:103-108

Kettunen E, Nicholson AG, Nagy B, Wikman H, Seppänen JK, Stjernvall T, Ollikainen T, Kinnula V, Nordling S, Hollmén J, Anttila S, Knuutila S (2005) L1CAM, INP10, P-cadherin, tPA and ITGB4 over-expression in malignant pleural mesotheliomas revealed by combined use of cDNA and tissue microarray. Carcinogenesis.26:17-25

Lalla RV, Boisoneau DS, Spiro JD, Kreutzer DL (2003) Expression of vascular endothelial growth factor receptors on tumor cells in head and neck squamous cell carcinoma. Arch Otolaryngol Head Neck Surg. 129:882-888

Legler DF, Loetscher M, Roos RS, Clark-Lewis I, Baggiolini M, Moser B (1998) B cell-attracting chemokine 1, a human CXC chemokine expressed in lymphoid tissues, selectively attracts B lymphocytes via BLR1/CXCR5. J Exp Med 187:655-660

Liu N, Zhang G, Bi F, Pan Y, Xue Y, Shi Y, Yao L, Zhao L, Zheng Y, Fan D (2007) RhoC is essential for the metastasis of gastric cancer. J Mol Med 85:1149-1156

Martinez I, Wang J, Hobson KF, Ferris RL, Khan SA (2007) Identification of differentially expressed genes in HPV-positive and HPV-negative oropharyngeal squamous cell carcinomas. Eur J Cancer 43:415-432

Mori M, Manuelli C, Pimpinelli N, Bianchi B, Orlando C, Mavilia C, Cappugi P, Maggi E, Giannotti B, Santucci M (2003) BCA-1, A B-cell chemoattractant signal, is constantly expressed in cutaneous lymphoproliferative B-cell disorders. Eur J Cancer 39:1625-1631

Nadal A, Jares P, Cazorla M, Fernández PL, Sanjuan X, Hernandez L, Pinyol M, Aldea M, Mallofré C, Muntané J, Traserra J, Campo E, Cardesa A (1997) p21WAF1/Cip1 expression is associated with cell differentiation but not with p53 mutations in squamous cell carcinomas of the larynx. J Pathol 183:156-163

Nakahara Y, Shiraishi T, Okamoto H, Mineta T, Oishi T, Sasaki K, Tabuchi K (2004) Detrended fluctuation analysis of genome-wide copy number profiles of glioblastomas using array-based comparative genomic hybridization. Neuro-oncol 6: 281–289

Noskov VN, Araki H, Sugino A (1998) The RFC2 gene, encoding the third-largest subunit of the replication factor C complex, is required for an S-phase checkpoint in Saccharomyces cerevisiae. Mol Cell Biol 18: 4914–4923

O'Brien SL, Fagan A, Fox EJ, Millikan RC, Culhane AC, Brennan DJ, McCann AH, Hegarty S, Moyna S, Duffy MJ, Higgins DG, Jirström K, Landberg G, Gallagher WM (2007) CENP-F expression is associated with poor prognosis and chromosomal instability in patients with primary breast cancer. Int J Cancer 120:1434-1443

Robinson PA, Markham AF, Schalkwijk J, High AS (1996) Increased elafin expression in cystic, dysplastic and neoplastic oral tissues. J Oral Pathol Med 25:135-139

Sha D, He YJ (2006) Expression and clinical significance of VEGF and its receptors Flt-1 and KDR in nasopharyngeal carcinoma. Ai Zheng 25:229-234

Speicher MR, Schoell B, du Manoir S, Schröck E, Ried T, Cremer T, Störkel S, Kovacs A, Kovacs G (1994) Specific loss of chromosomes 1, 2, 6, 10, 13, 17, and 21 in chromophobe renal cell carcinomas revealed by comparative genomic hybridization. Am J Pathol 145: 356–364

Spiro RH, Guillamondegui Jr O, Paulino AF, Huvos AG (1999) Pattern of invasion and margin assessment in patients with oral tongue cancer. Head Neck 21:408–413

Yang XL, Chu YW, Xu L, Li BH, Xiong SD (2006) Inoculation of the cells transfected with IP10 inhibited experimental tumor metastases in vivo. Zhonghua Yi Xue Za Zhi. 86:176-81

Zhang R, Zhang H, Zhu W, Pardee AB, Coffey RJ Jr, Liang P (1997) Mob-1, a Ras target gene, is overexpressed in colorectal cancer. Oncogene 14: 1607-1610

Biomarkers for Clinical Oncology

Yasutsuna Sasaki

Department of Medical Oncology, Saitama International Medical Center – Comprehensive Cancer Center, Saitama Medical University, 1397-1 Yamane, Hidaka, Saitama 350-1298, Japan

Summary. The concept of biomarker to predict not only drug induced toxicities but also clinical response and prognosis of the patients is extremely important in medical oncology. Among many kinds of biomarkers or candidates of biomarker, both pharmacokinetic dose adjustment by analyzing polymorphism of drug metabolizing enzyme and responder enrichment based on pharmacodynamic biomarkers are essential approach for individualized dosing of anticancer agents. Dose modification strategy based on pharmacogenetic- pharmacokinetic information to minimize life-threatening toxicities is extremely important approach for conventional cytotoxic agents. Recent studies reported polymorphism of drug-metabolizing enzymes in anticancer agents such as tegafur (FT), letrozole, tamoxifen, 6-merucaptopurine, 5-fluorouracil (5-FU) and irinotecan. Especially both irinotecan and S1 (tegafur/CDHP/oxonic acid) are key drugs and widely prescribed in Japan in the treatment of gastrointestinal cancers. Genetic polymorphisms of UGT1A1, a crucial drug-metabolizing enzyme of irinotecan, are essential determinants of individual variation in susceptibility to irinotecan induced toxicities. In addition to UGT1A1*28, both UGT1A1*6 and UGT1A1*27 polymorphism increase the risk of toxicities. Establishment of optimal dose of irinotecan for the patients with these polymorphisms is warranted. On the other hand, to analyze contributions of CYP2A6 genotype, plasma 5-chloro-2,4-dihydroxipyridine (CDHP) levels to the pharmacokinetics of FT and 5 -FU is also important issue not only to optimize S1 administration but also to understand possible ethnic difference of this drug. Our prospective clinical trial concludes that clearance of FT is associated only with CYP2A6 genotype. The clearance of FT seen in patients with one- and two-variant alleles is significantly lower than that seen in wild type, respectively. The AUC0-8 for 5-FU correlated with exposure of CDHP and creatinine clearance. These basic information can apply to clinical practice at bedsides.

Genome-Wide cDNA Microarray Screening of Gene Expression Profiles Correlated with Resistance to Anti-Cancer Drug Treatment and Radiation in Oral Squamous Cell Carcinoma

Satoru Shintani

Department of Oral and Maxillofacial Surgery, Showa University School of Dentistry

Summary. Chemotherapy and radiation therapy are an acceptable treatment modality for patients with oral squamous cell carcinomas (OSCC). However, the response of carcinoma to radiation and/or chemotherapy varies in each patient. To choose the proper therapy as well as to avoid untoward side effects, a useful method to predicting the effectiveness of radiation and/or chemotherapy for individual patient must be established. To identify genes correlating with sensitivity to anti-cancer drug and radiation in oral cancer treatment, we performed cDNA microarray analysis of OSCC cell lines. In OSCC cell lines, we compared the gene expression profiles with the sensitivities, that were measured by CD-DST method, to five anti-cancer drugs and with the radiation sensitivity assessed by a standard colony formation assay. In this study, we found significant associations between dozens of gene expression levels and resistance to anti-cancer drugs and radiation.

Key word. gene expression profiles, cDNA microarray, anti-cancer drug, radiation, oral squamous cell carcinoma

Introduction

Chemotherapy and radiation therapy play an important role in OSCC treatment. They allow to improve the overall survival rates and to maintain oral morphology and its important function. However, the response of carcinoma to these therapies varies in each patient, and no survival benefit has been observed in non-responders (Andreadis 2003). Difference among patients with respect to the effectiveness of anti-cancer drugs and/or radiation has been associated with variation in gene expression profiles in cancer cells (Golub 1999, Goliub 2001).

Using a cDNA microarray analysis, we obtained comprehensive gene-expression profiles of OSCC cell lines. In present study, we identified the candidates of molecular markers for the potential of predicting sensitivity to anti-cancer drugs and radiation therapy.

Materials and Methods

Evaluation of sensitivity to anti-cancer drugs using CD-DST

The prepared tumor cell suspension was added to a collagen solution (Collagen Gel Culture Kit, Nitta Gelatin) and allowed to gel at 37°C in a CO2 incubator. We introduced each of five anti-cancer drugs into the wells. After the incubation for 7 days, Neutral red was added and the cells were fixed by 10% neutral formalin buffer and then quantified by an image analysis.

Clonogenic survival assay for radiation sensitivity

OSCC cells were plated and, 48 hr later, they were irradiated with various single radiation doses by a linear accelerator (MBR.1520R, Hitachi Medico, Tokyo), and then incubated for approximately 10 days to allow form to colonies. The survival fractions were calculated as a ratio of plating efficiencies in treated and untreated cells.

cDNA Microarrays

Applied Biosystems Human Genome Survey Arrays were used to analyze the transcriptional profiles of RNA samples of OSCC cell lines. Array hybridization (two arrays per sample), Chemiluminescence detection, image acquisition and analysis were performed using Applied Biosystems Chemiluminescence Detection Kit and Applied Biosystems 1700 Chemiluminescent Microarray Analyzer following manufacturer's protocol.

Identification of the genes associated with resistance to anti-cancer drug and radiation

Gene spring 7 software (Silicon Genetics) was used to extract assay signal and noise ratio values from the microarray images. To estimate correlation between the expression ratio and sensitivity to each drug, we calculated a Pearson correlation coefficient by the following formula: $r = \Sigma(x_i - x_{mean})(y_i - y_{mean}) / \sqrt{\Sigma(x_i - x_{mean})^2 (y_i - y_{mean})^2}$, and we selected genes showing significant correlation ($P < 0.01$).

Results

Genes associated with sensitivity to anti-cancer drug

Results of CD-DST experiments for five anti-cancer drugs were obtained and Pearson's correlation coefficients between gene-expression levels and chemosensitivity of 9 OSCCs to each of the five drugs were calculated (representative diagrams shown in Figure 1). We found the dozens of genes were associated with chemosensitivity (P<0.01).

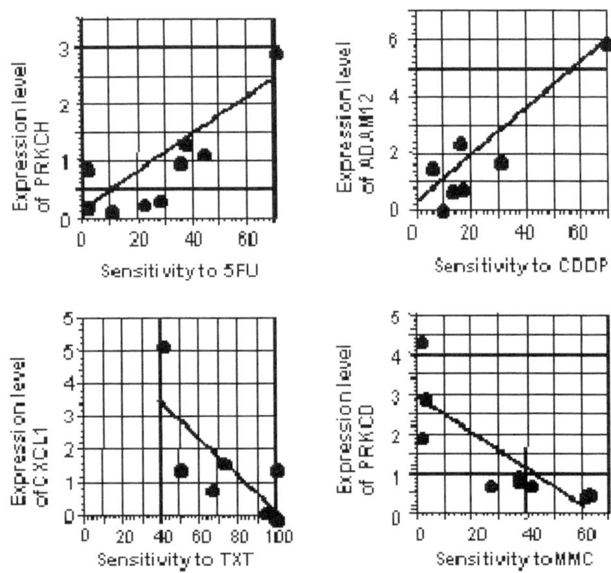

Fig. 1 Chemosensitivities are plotted on the X-axes, and expression levels of the genes are plotted on the Y-axes. Expression levels of PRKCH in the 9 OSCC cell lines show positive association with sensitivity to 5FU (r2=0.701, P<0.01), expression levels of ADAM12 show positive association with sensitivity to CDDP (r2=0.855, p<0.01), expression levels of CXCL1 show negative association with sensitivity to TXT (r2=0.756, p<0.01), and expression levels of PRKCD show negative association with sensitivity to MMC (r2=0.677, p<0.01).

Clonogenic survival of seven OSCC cell lines after radiation

To determine the radiation sensitivity, a clonogenic survival assay was performed on the seven human OSCC cell lines. Ca9-22 and SCC25 were more radiosensitive than the other cells. SCC111 and HSC2 did not respond to radiation.

Genes associated with sensitivity to radiation

To discover genes that might be associated with radiation resistance, we then compared expression levels with radiation sensitivity measured by clonogenic survival assay. Pearson's correlation coefficients between gene-expression levels and radiation sensitivity of 7 OSCCs were calculated. We found the dozens of genes to be associated with radiation resistance (P<0.01).

The relationship between gene expression levels of FGF2, Fli-1, and G1P3 and radiation sensitivity

The correlation between the D10 of these SCC cells and the FGF2, Fli-1, and G1P3 genes expression levels was determined. As shown in Figure 2, there was a significant correlation between the magnitude of those gene expressions and SCC cells radiosensitivity (FGF2: R2 = 0.809, G1P3: R2=0.716, Fli-1:R2=0.750).

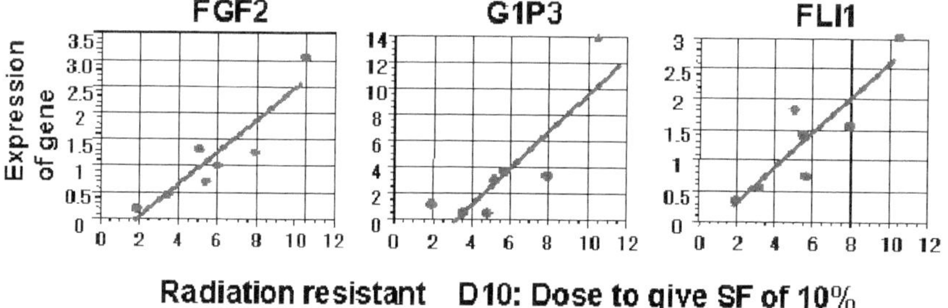

Fig. 2 Relationship between FGF2, G1P2, and Fli1 expression and radiation sensitivity (D10) of oral squamous cell carcinoma cell lines.

The relationship between FGF2, Fli-1, G1P3 expression and response to radiation

The correlation between FGF2, Fli-1 and G1P3 expression and the pathologic response to radiation are shown in Fig. 3 and Table 1. A statistically significant association was found between those proteins expression and the pathologic responses when they were categorized into the low and high groups (FGF2: P = 0.0011, G1P3: P=0.0487, Fli1: P=0.0492).

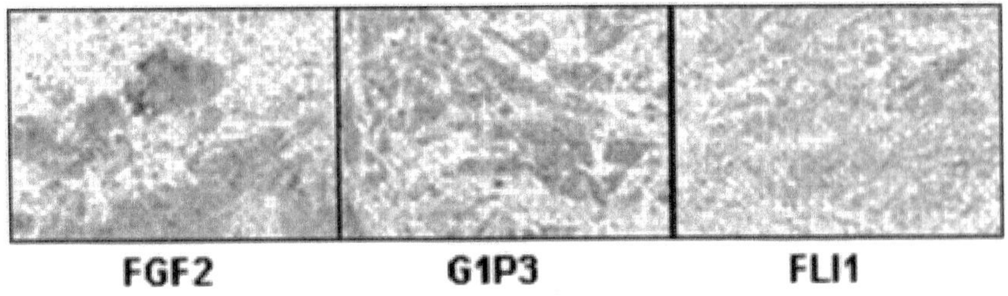

Fig. 3 FGF2, Fli-1, G1P3 proteins were localized to the cytoplasm in tumor specimens from oral SCC patients before preoperative radiation therapy. Refer to color plates.

Immunohistochemical staining of FGF2 product before radiation therapy	Pathological response 0 - 2+	Pathological response 3+ - 4+	
0 -1+	6	7	P=0.0011
2+ - 3+	12	1	

Immunohistochemical staining of FLI1 product before radiation therapy	Pathological response 0 - 2+	Pathological response 3+ - 4+	
0 -1+	6	6	P=0.0487
2+ - 3+	12	2	

Immunohistochemical staining of G1P3 product before radiation therapy	Pathological response 0 - 2+	Pathological response 3+ - 4+	
0 -1+	4	5	P=0.0492
2+ - 3+	14	3	

Table 1 Correlation between pathologic response and FGF2, Fli-1, and G1P3 expressions in OSCC specimens

Discussion

Our study investigated the molecular patterns of genes expression that contribute to resistance to anti-cancer drug and radiation using a microarray system. To identify the molecular markers that might predict resistance to anti-cancer drugs and radiation of individual tumors, we analyzed gene expression profiles of OSCC cell lines. The cDNA microarray method is now widely used to analyze expression of thousands of genes simultaneously in cancer cells. cDNA microarray analysis for the purpose of screening for biomarkers may be useful to predict the outcome of therapy for OSCC. Through a computational analysis, we selected some candidate genes that are likely to be associated with resistance to anti-cancer drug. The expressions of protein kinase C (PKC) isoforms, delta and eta, are associated with the resistance to 5FU and MMC in our experiments. According to a recent report (Carter 2004), some natural compounds and pharmacologic analogues that modulate PKC activity can be used effectively to prevent or treat cancer. The other candidates, such as ADAM12 and CXCL1, are also associated with resistance to CDDP and TXT. A recent paper (Kveiborg 2005) suggested that the ADAM12 is able to have influence on apoptosis and it may contribute to tumor progression. On the other hand, fibroblast growth factor 2 (FGF2), friend leukemia inserton -1 (Fli-1) and an interferon inducible gene 6-16 (G1P3) are associated with the resistance to radiation. According to a recent report (Griffin 2002), pharmacologic analogues that inhibit FGF receptor tyrosine kinase activity can enhance the tumor radiation response. These data should give useful information contributing to identifying predictive markers for therapeutic sensitivity. It may eventually provide "personalized chemotherapy" for individual patient, and may lead to the development of alternative treatment modalities for chemotherapy and/or radiation therapy.

References

Andreadis C, Vahtsevanos K, Sidiras T, Thomaidis I, Antoniadis K, Mouratidou D (2003) 5-Fluorouracil and cisplatin in the treatment of advanced oral cancer. Oral Oncol. 39: 380-385.

Carter CA, Kane CJ (2004) Therapeutic potential of natural compounds that regulate the activity of protein kinase C. Curr Med Chem 11: 2883-2902.

Golub TR, Slomin DK, Tamayo P, Huard C, Gaasenbeek M, Masirov JP, Coller H. Loh ML, Downing JR, Caligiuri MA, Bloomfield CD and Lander ES (1999) Molecular classification of cancer: class discovery and class prediction by gene expression monitoring. Science 286: 531-537, 1999.

Golub TR(2001) Genome-wide views of cancer. N Engl J Med 344: 601-602.

Griffin RJ, Williams BW, Wild R, Cherrington JM, Park H, Song CW (2002).Simultaneous inhibition of the receptor kinase activity of vascular endothelial, fibroblast, and platelet-derived growth factors suppresses tumor growth and enhances tumor radiation response. Cancer Res. 62(6):1702-1706.

Kveiborg M, Frohlich C, Albrechtsen R, Tischler V, Dietrich N, Holck P, Kronqvist P, Rank F, Mercurio AM,ewer UM (2005) A role for ADAM12 in breast tumor progression and stromal cell apoptosis. Cancer Res 65:4754-4761.

Part IV
Poster Session

Drug Accumulation and Efflux Do not Contribute to Acquired Gefitinib Resistance

Sojiro Kusumoto[1], Tomohide Sugiyama[1], Masanao Nakashima[1], Toshimitsu Yamaoka[1], Takashi Hirose[1], Tsukasa Ohnishi [1], Naoya Horichi[1], Mitsuru Adachi[1], Kazuto Nishio[3] Nagahiro Saijo[4], Toshio Kuroki[5], Tohru Ohmori[2]

[1]First Department of Internal Medicine, Showa University School of Medicine, 1-5-8 Hatanodai, Shinagawa-ku, Tokyo, 142-8666, Japan
[2]Institute of Molecular Oncology, Showa University, 1-5-8 Hatanodai, Shinagawa-ku, Tokyo, 142-8555, Japan
[3]Institute of Molecular Biology, Kinki University 377-2 Ohnohigashi Osakasayama-shi, Osaka-fu, 589-8511, Japan
[4]National Cancer Center Research Institute, Internal Medicine, National Cancer Hospital East, 6-5-1 Kashino-ha Kashiwa-shi Chiba, 277-8577, Japan
[5]Gifu University, 1-1 Yanagido Gifu-shi, Gifu 501-1194, Japan

Summary. Gefitinib is one of tyrosine kinase inhibitors and often has dramatic effect on non-small-cell lung cancer (NSCLC) expressed mutant EGFR. However, most of the patients who respond to gefitinib eventually experience tumor recurrence. To clarify the mechanism of this acquired resistance, we examined cellular accumulation and efflux of gefitinib using gefitinib sensitive and resistant NSCLC cells. We used four NSCLC cell lines; PC-9: hypersensitive to gefitinib, expressed a 15 bp deletion mutant EGFR, PC-9/ZD2001 and PC-9/ZD1k1: acquired-resistant to gefitinib, PC-9/ZD2001R: a revertant reacquired sensitivity to gefitinib. To measure the cellular accumulation, cells were exposed to 1 μM of [^{14}C]gefitinib for 3 to 30 min at 37 °C For the measuring of drug efflux, cells were exposed gefitinib for 30 min, and then cells were washed and further incubated in drug free medium. After the incubation, cells were lysed and the radioactivities were counted. There was no significant difference of gefitinib accumulation in those cell lines (range 642-731μmol/g protein at 30 min). The efficiencies of gefitinib-efflux were almost the same in those cell lines (about 80% of gefitinb was discharged at 15 min). According these findings, we demonstrated that acquired resistance to gefitinib did not depend on cellular accumulation or efflux of this drug.

Key words. accumulation, efflux, gefitinib, resistance

Introduction

Geftinib was developed as an Epidermal Growth Factor Receptor (EGFR) tyrosine kinase inhibitor (TKI) and has been approved for treatment to advanced non small cell lung cancer (NSCLC). This agent frequently shows dramatic response to NSCLCs expressed mutant EGFR. However, most of the patients who respond to gefitinib eventually experience tumor progression. To evaluate this acquired resistance we previously established a gefitinib resistant NSCLC cell line, and have been exploring the mechanisms. This time we hypothesized that cellular pharmacokinetics contributed to the drug resistance, then examined drug accumulation and efflux on gefitinib-sensitive and resistant NSCLC cell lines.

Materials and Methods

Chemicals and cell lines

Radio-labeled[^{14}C]gefitinib was generously provided by AstraZeneca Pharmaceuticals. We used four human NSCLC cell lines; PC-9 expressed a 15 bp deletion mutant EGFR, is hypersensitive to gefitinib. PC-9/ZD2001 and PC-9/ZD1k1 are acquired-resistant to gefitinib, were derived from PC-9 through continuous exposure of this drug PC-9/ZD2001 cells and PC-9/ZD1k1 cells could survive on 200nM, 1μM of gefitinib respectively. PC-9/ZD2001R: a revertant reacquired sensitivity to gefitinib. These cells were cultured on RPMI 1640 medium added 10% fetal bovine serum, and were kept viable in 5%CO2 incubator at 37 °C.

Growth inhibition assay

As MTT assay, we evaluated the cytotoxic effects of gefitinib to NSCLC cell lines.

Accumulation and efflux assay

Cells were adjusted to 2.0-3.0×10^5/well. To measure the cellular drug accumulation, we added [^{14}C]gefitinib to each well, adjusting final concentration to 1μM. After exposing the cells to 1 μM of [^{14}C]gefitinib for 3 to 30 min at 37 °C, we rapidly washed the cells with ice cold PBS, then lysed them and the radioactivities of lysate were counted by liquid scintillation counter. For the measuring of drug efflux, cells were exposed to 1 μM of [^{14}C]gefitinib for 30 min, and then cells were washed with ice cold PBS and further incubated in drug free medium for 3 to 15min at 37 °C. Subsequently washing the cells again, we lysed them and counted their radioactivities. Measuring absorbances at 570nm with microplate reader, we counted protein concentrations of each well. After collecting these data, we calculated accumulation and efflux of gefitinib per protein (μmol/g).

Results

We measured cytotoxic effect of gefitinib in gefitinib-sensitive, resistant and revertant NSCLC cells by MTT assay. (Table1) Gefitinib-resistant PC-9/ZD2001, PC-9/ZD1k1 cells showed approximately 4.0-fold, 5.5-fold higher resistance to gefitinib than their parental cells. In PC-9/ZD2001R cells, the sensitivity was completely restored.

We hypothesized that cellular pharmacokinetics was responsible to sensitivities of gefitinib, and then measured drug accumulation and efflux on these NSCLC cells using radio-labeled [^{14}C]gefitinib. However, there was no significant difference of drug accumulation in those cell lines (range 642-731μmol/g protein at 30 min). The efficiencies of gefitinib-efflux were almost the same in those cell lines (about 80% of gefitinb was discharged at 15 min) (Fig.1) Our results demonstrated that acquired resistance to gefitinib did not depend on cellular accumulation and efflux of this drug.

NSCLC cell lines	IC40 values
PC-9	53.0 ± 8.1 nmol/L
PC-9/ZD2001	211.1 ± 32.4 nmol/L
PC-9/ZD1k1	291.5 ± 15.5 nmol/L
PC-9/ZD2001R	46.3 ±10.2 nmol/L

Table 1. Cytotoxic effects of gefitinib in NSCLC cell lines

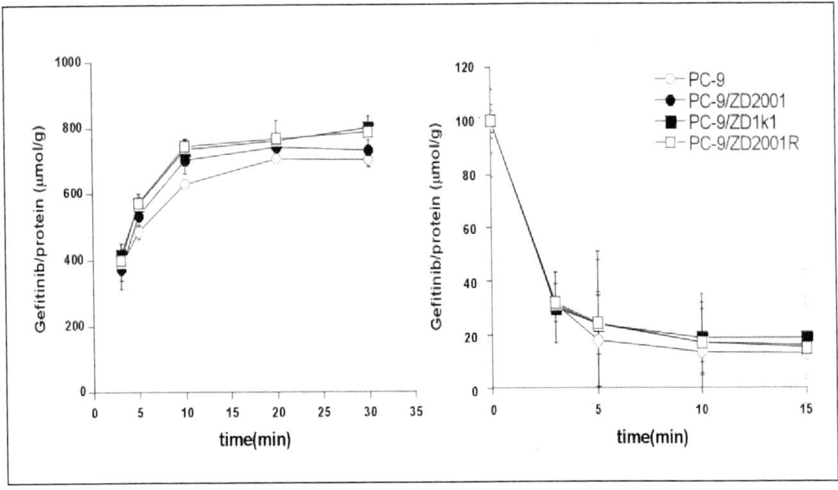

Fig.1. There was no significant difference of drug accumulation in four NSCLC cell lines (range 642-731μmol/g protein at 30 min). The efficiencies of gefitinib-efflux were almost the same in those cell lines (about 80% of gefitinb was discharged at 15 min)

Discussion

To elucidate gefitinib resistance, we focused on cellular accumulation and efflux, because drug resistant cancer cells occasionally expressed ATP binding cassette (ABC) transporters that worked as pump to extrude drugs(Gottesman et al. 2002). Breast cancer resistance protein (BCRP) / ABCG2 is an ABC transporter, and several reports indicated that BCRP should be related to gefitinib resistance(Ozvegy-Laczka, Hegedus, Varady, et al. 2004) (Yanase 2004) (Usuda 2007). Their speculations about the mechanism of geifitnib resistance were also that BCRP might be active as efflux pump of this drug. Gefitinib-resistant PC-9/ZD2001 cells expressed mRNA of BCRP approximately 3-fold higher than gefitinib-sensitive PC-9. (data not shown) However, our results did not show any significant difference in these cells upon drug accumulation and efflux. In addition, other investigators also reported that expression of BCRP did not alter uptake or efflux of geifitinib (Stewart et al. 2004). Therefore, we demonstrate that BCRP should not be efflux pump of gefitinib.

Although we examined only cellular accumulation and efflux of gefitinib, the intracellular movement of this drug has been unknown. We think that further examinations of the movement should be demanded as the next stage.

References

Gottesman MM, Fojo T, Bates SE. Multidrug resistance in cancer: role of ATP-dependent transporters. Nat Rev Cancer 2002;2(1):48-58.

Ozvegy-Laczka C, Hegedus T, Varady G, et al. High-affinity interaction of tyrosine kinase inhibitors with the ABCG2 multidrug transporter. Mol Pharmacol 2004;65(6):1485-95.

Stewart CF, Leggas M, Schuetz JD, et al. Gefitinib enhances the antitumor activity and oral bioavailability of irinotecan in mice. Cancer Res 2004;64(20):7491-9.

Usuda J, Ohira T, Suga Y, et al. Breast cancer resistance protein (BCRP) affected acquired resistance to gefitinib in a "never-smoked" female patient with advanced non-small cell lung cancer. Lung Cancer 2007;58(2):296-9.

Yanase K, Tsukahara S, Asada S, Ishikawa E, Imai Y, Sugimoto Y. Gefitinib reverses breast cancer resistance protein-mediated drug resistance. Mol Cancer Ther 2004;3(9):1119-25.

Difference of EGFR-Binding Proteins between Wild Type and Mutant EGFR

Tsuyoki Kadofuku[1], Tomoko Kanome[1], Kazuto Nishio[3], Toshio Kuroki[4], and Tohru Ohmori[1,2]

[1]Institute of Molecular Oncology, Showa University, [2]First Department of Internal Medicine, Showa University School of Medicine, 1-5-8 Hatanodai, Shinagawa-ku, Tokyo 142-8555, Japan, [3]Department of Genome Biology, Kinki University School of Medicine, 377-2 Osakasayama, Osaka 589-8511, Japan, [4]Gifu University, 1-1 Yanagido, Gifu 501-1193, Japan

Summary. Non-small cell lung cancer cells expressed mutant EGFR are more sensitive to gefitinib (Iressa) than that expressed wild type EGFR. To elucidate the mechanism of the hypersensitivity to gefitinib in the mutant EGFR, we explored the difference of EGFR-binding proteins between wild type and a 15 bp deletion mutant EGFR using respective stable transfectant cells. EGFR-binding proteins in respective transfectant cells were collected by co-precipitation with polyclonal anti-. EGFR and the co-precipitate proteins were separated by two-dimensional polyacrylamide gel electrophoresis (2D-PAGE). Both co-precipitate proteins showed a similar 2D-PAGE pattern, however, several proteins differed between both transfectant cells. Among these proteins, one protein, detected to high concentration in mutant EGFR tranfectant cells, was identified as heat shock protein 70 (HSP 70) by peptide mass finger printing. Much binding of HSP 70 to mutant EGFR was also confirmed by Western blotting. There was no significant difference of HSP 70 protein expression between both transfectant cells. These results suggest that the difference of HSP70 binding to EGFR may modify the EGFR-downstream signaling and influence the sensitivity to gefitinib.

Key words. Gefitinib, non-small cell lung cancer cells, EGFR-binding proteins, heat shock protein 70, two-dimensional polyacrylamide gel electrophoresis, LC-MS/MS

1 Introduction

Epidermal growth factor receptor (EGFR), a 170-kDa membrane glycoprotein with tyrosine kinase activity, plays central roles in cell proliferation, survival, migration, differentiation, and angiogenesis.

EGFR-mediated signaling is thought to play an important role in the progression of epithelial neoplasm. Increased EGFR expression has been reported in a wide variety of human tumors. Gefitinib (Iressa) is an orally active EGFR-tyrosine kinase inhibitor that block signal transduction pathways in cancer cell proliferation, survival and other host-dependent process that promote cancer growth. Gefitinib has demonstrated antitumor efficacy in patients with relapsed or recurrent non-small cell lung cancer (NSCLC) and has received approval for the treatment of advanced NSCLC (Herbst 2003, Fukuoka et al. 2003, Kris et al. 2003). A possible paradigm for determining response to gefitinib has been reported that in-frame mutation of EGFR are well-corrected to the hyper-responsiveness to gefitinib in patients with NSCLC (Lynch et al. 2004, Paez et al. 2004). These mutations were small, in-frame deletion or substitutions clustered around the ATP-binding site in exons 18, 19, and 21 of EGFR.

We previously identified a 15-bp in-frame deletion in exon 19 of EGFR (2411-2425) in a gefitinib-hypersensitive NSCLC cell line, PC-9. The PC-9 cell line was hypersensitive to gefitinib (IC_{40} =53.0±8.1 nM) as compared with another NSCLC cell line, PC-14, which expresses wild type EGFR (IC_{40} =47.0±9.1 μM). These observations suggest that this deletion mutant might be an active mutant EGFR that may correlate with tumor responsiveness to gefitinib. In the present study, to elucidate the mechanism of the hypersensitivity to gefitinib in the mutant EGFR, we explored the difference of EGFR-binding proteins between wild type EGFR and a 15 bp deletion mutant EGFR using respective stable transfectant cells.

2 Materials and methods

2.1 Transfectant cells

Wild type EGFR stable transfectant cells (293_pEGFR) and a 15 bp deletion mutant EGFR stable transfectant cells (293_pΔ15) were kindly provided by Dr. Fukumoto (Shien Laboratory, National Cancer Center) and maintained in RPMI 1640 medium containing 10% fetal calf serum (FCS) at 5% CO_2. 293_pΔ15 cell is about 1000-fold sensitive to gefitinib as compared to 293_pEGFR cell.

2.2 EGFR-stimulation and sample preparation

To eliminate the effect of FCS, cells were starved by 0.1% FCS contained medium for overnight, then exposed to 10 ng/ml of TGFα for 1 h. After the exposure, cells were washed 3 times in 50 ml of cold PBS, lysed with

KLB buffer (50 mM Tris-HCl, pH 7.4, 150 mM NaCl, 0.1% Triton X-100, 0.1% NP-40, and 4 mM EDTA-2Na) containing protease and phosphatase inhibitors (20 mM NaF, 1 mM H_3VO_4, 10 μg/ml leupeptin, 10 μg/ml aprotinin, and 10 μg/ml PMSF), and centrifuged at 15,000 rpm for 15 min. The supernatant was mixed with an EGFR antibody (EGFR1005 rabbit polyclonal antibody, Santa Cruz Biotechnology) for 1 hr and the EGFR-bound proteins were co-precipitated by Protein A-Sepharose beads. After adequate washing, the EGFR-bound proteins were eluted by 10 mM sodium citrate (pH3.0) and collected by acetone precipitation (3 vol. of cold acetone).

2.3 Two-dimensional polyacrylamide gel electrophoresis

Two-dimensional polyacrylamide gel electrophoresis (2D-PAGE) was performed as follows. First dimension of 2D-PAGE was done on non-linear pH 3-10 Immobiline™ DryStrip (7 cm long) using Ettan IPGphor (GE healthcare, Uppsala, Sweden). Protein samples were dissolved in 125 μl of a sample buffer (8M urea, 4% CHAPS, 40 mM dithiothreitol (DTT), and 0.5% IPG buffer). After rehydration for 12 hrs at 20℃, electrophoresis was run at 300V for 3 hrs (linear), 1000V for 30 min (linear), 5000V for 1.5 hrs (linear), finally 5000V for 30 min (one step) at 20℃. Second dimension was performed on 10% SDS-PAGE (14 cm width, 7 cm height, 0.1 cm thickness). Before the second dimension, Immobiline strips were equilibrated in 50 mM Tris-HCl (pH 8.8) containing 6M urea, 30% glycerol, 2% SDS, and 1% DTT for 20 min, and sealed with a melted 0.5% agarose solution (50 mM Tris-192 mM glycine and 0.1% SDS) on top of the second dimension gel. Electrophoresis was run at 20 mA constant current for 2 hrs at 20℃. Proteins were visualized by a Silver Staining Kit (PlusOne™, GE healthcare). Proteins for mass analysis were stained without glutardialdehyde.

2.4 Mass analysis and peptide mass finger printing

In-gel digestion was performed according to the method of Shevchenko et al. (Shevchenko et al. 1996) with some modifications. Gel pieces containing silver-stained spots (stained without glutardialdehyde) were excised and washed with 15 mM potassium ferricyanide-50 mM sodium thiosulfate. After washing in distilled water, the gel pieces were shrank by dehydration in 100% acetonitrile, which was then removed, they were dried in a vacuum centrifuge. A volume of 10 mM DTT in 100 mM NH_4HCO_3 sufficient to cover the gel pieces was added, and the proteins were reduced for 1 h at 60℃. The DTT solution was replaced with the same volume of 50 mM iodoacetamide in 100mM NH_4HCO_3. After 30 min incubation in the dark at room temperature, the gel pieces were washed with 100mM NH_4HCO_3, dehydrated by addition of 100% acetonitrile and dried in a vacuum centrifuge. To the gel pieces were swollen in 20 μl of 50 mM NH_4HCO_3 containing 50 ng of trypsin and kept at 37℃ for overnight. Peptides were extracted by one change of 50 mM NH_4HCO_3 and three changes of 5% formic acid in 50% acetonitrile and dried in a vacuum centrifuge. The peptides were dissolved in 10 μl of 0.1% trifluoroacetic acid (TFA) and desalted using ZipTip (Millipore Corp.). The desalted-concentrated peptides were eluted 3 times with 2 μl of 50% acetonitrile containing 0.1% TFA and analyzed on a LC-MS/MS (Q-Tof micro™). Peptide mass finger printing data were searched using Mascot search algorithm (Matrix Science).

2.5 Western blotting

After 10% SDS-PAGE, proteins were transferred to a nitrocellulose membrane (Hybond-C, GE healthcare) at 2 mA/cm^2 for 2 hrs. Anti-EGFR (1005) rabbit polyclonal IgG and anti-HSP 70 (W27) mouse monoclonal IgG_{2a} (both Santa Cruz Biotechnology) were used as primary antibodies. Proteins were detected using Western Lighting™ Chemiluminescence Reagent Plus (Perkinelmer corp.).

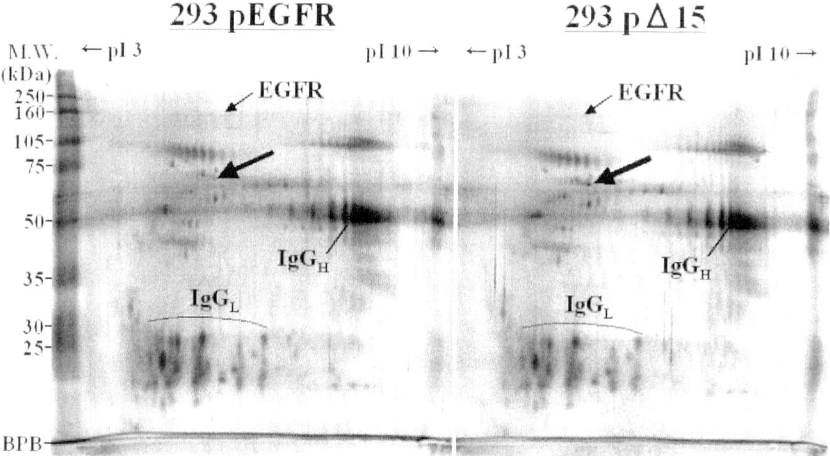

Fig. 1 2D-PAGE of EGFR-bound proteins. Polyclonal anti-EGFR (40 μg) was mixed with respective cell lysate and co-precipitate proteins were separated by 2D-PAGE. Proteins were visualized with silver staining. Bold arrows show identified protein in this study. IgG_H: immunoglobulin G heavy chain, IgG_L: immunoglobulin G light chain. Refer to color plates.

3 Results and discussion

To find the difference of EGFR-binding proteins between wild type and mutant EGFR, we mixed respective cell lysate and polyclonal anti-EGFR, and co-precipitate proteins were analyzed by 2D-PAGE. In Fig.2, 2D-PAGE patterns of respective co-precipitate proteins are shown. Co-precipitate proteins showed a similar 2D-PAGE pattern, however, several proteins differed between 293_pEGFR and 293_pΔ15 cells (Fig. 1). We tried identification of these different proteins by peptide mass fingerprinting (PMF). Silver-stained proteins in 2D-PAGE gel were excised and digested with trypsin as described in Section 2.4. Peptide masses were analyzed by LC-MS/MS and the peptide mass data were searched using a Mascot search algorithm. As a result, one protein (indicated by arrow in Fig. 1, detected to high concentration in 293_pΔ15 cells) was identified as heat shock protein 70 (HSP 70) (Mowse score 494, peptides matched 15 and sequence coverage 32%).

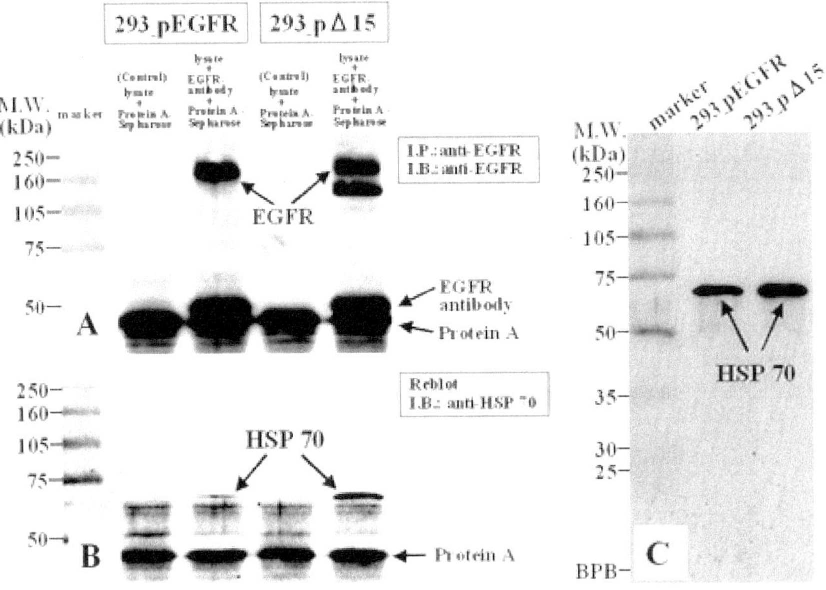

Fig. 2 Western blot analysis of EGFR-bound proteins and expression of HSP 70 protein in transfectant cells. A and B; each cell lysate was mixed with or without (control) polyclonal anti-EGFR (1 μg). Co-precipitate proteins were separated by 10% SDS-PAGE, transferred to a nitrocellulose membrane, and detected with anti-EGFR (A) and anti-HSP 70 (B), respectively. C; an equivalent volume of cell lysate (1 μg proteins) was separated by 10% SDS-PAGE and detected with anti-HSP 70.

To examine the difference of HSP 70 binding activity between wild type and mutant EGFR, an equal amount of polyclonal anti-EGFR (1 μg) was mixed with respective cell lysates and co-precipitate proteins were analyzed by Western blotting. Figs. 2A and 2B are Western blots of co-precipitate proteins, detected with anti-EGFR (Fig. 2A) and with anti-HSP 70 (Fig. 2B), respectively. As shown in Fig.1A, EGFR was detected with almost the same intensity, but the intensity of HSP 70 was obviously high in 293_pΔ15 transfectant cells (Fig. 2B). This difference may depend on HSP 70 expression level in transfectant cells. Therefore, we compared HSP 70 level between 293_pEGFR and 293_pΔ15 transfectant cells. However, no significant differences were observed between both transfectant cells (Fig. 2C). These observations clearly show that mutant EGFR has high HSP 70 binding activity in comparison with wild type EGFR. In 293_pΔ15 transfectant cell, EGFR was observed as double bands on Western blot (Fig. 2A). The reason why mutant EGFR was detected as double bands is unclear. Judging from their molecular weights, lower band may be degradation products occurred during experiments.

In the present study, we examined the difference between wild type EGFR and a 15 bp deletion mutant EGFR and detected HSP 70 as a candidate protein to influence gefitinib sensitivity. HSP70 is known as not only a chaperone but also a transactivator of EGFR under the heat-stress (Evdonin et al. 2006). With these reports, our results suggest that there is possibility that the hyper HSP70 binding activity to mutant EGFR modulate the EGFR activity and EGFR-downstream signaling and may influence the sensitivity to gefitinib.

References

Herbst RD (2003) Dose-comparative monotherapy trials of ZD1839 in previously treated non-small cell lung cancer patients. Semin Oncol 30(*1 Suppl 1*):30-38.

Fukuoka M, Yano S, Giaccone G, Tamura T, Nakagawa K, Douillard JY, Nishiwaki Y, Vansteenkiste J, Kudoh S, Rischin D, Eek R, Horai T, Noda K, Takata I, Smit E, Averbuch S, Macleod A, Feyereislova A, Dong RP, Baselga J(2003)Multi-institutional randomized phase II trial of gefitinib for previously treated patients with advanced non-small cell lung cancer (The IDEAL 1 Trial) [corrected]. J Clin Oncol 21(12):2237-2246.

Kris MG, Natale RB, Herbst RS, Lynch TJ Jr, Prager D, Belani CP, Schiller JH, Kelly K, Spiridonidis H, Sandler A, Albain KS, Cella D, Wolf MK, Averbuch SD, Ochs JJ, Kay AC (2003) Efficacy of gefitinib, an inhibitor of the epidermal growth factor receptor tyrosine kinase, in symptomatic patients with non-small-cell lung cancer: a randomized trial. Jama 290(16):2149-2158.

Lynch TJ, Bell DW, Sordella R, Gurubhagavatula S, Okimoto RA, Brannigan BW, Harris PL, Haserlat SM, Supko JG, Haluska FG, Louis DN, Christiani DC, Settleman J, Haber DA (2004) Activating mutations in the epidermal growth factor receptor underlying responsiveness of non-small-cell lung cancer to gefitinib. N Engl J Med 350(21): 2129-2139.

Paez JG, Janne PA, Lee JC, Tracy S, Greulich H, Gabriel S, Herman P, Kaye FJ, Lindeman N, Boggon TJ, Naoki K, Sasaki H, Fujii Y, Eck MJ, Sellers WR, Johnson BE, Meyerson M (2004) EGFR mutations in lung cancer: correlation with clinical response to gefitinib therapy. Science 304(5676): 1497-1500.

Shevchenko A, Wilm M, Vorm O, and Mann M (1996) Mass spectrometric sequencing of proteins from silver-stained polyacrylamide gels. Anal Chem 68:850-858.

Evdonin AL, Guzhova IV, Margulis BA, Medvedeva ND (2006) Extracellular heat shock protein 70 mediates heat stress-induced epidermal growth factor receptor transactivation in A431 carcinoma cells. FEBS Lett 580(28-29):6674-6678.

Competitive Nuclear Export of Cyclin D1 and Hic-5 Regulates Anchorage-Dependence of Cell Growth and Survival

Kazunori Mori, Yukiko Oshima, Kiyoshi Nose and Motoko Shibanuma

Department of Microbiology, Showa University School of Pharmaceutical Sciences, 1-5-8 Hatanodai, Shinagawa-ku, Tokyo 142-8555, Japan

Summary. Cyclin D1 is a proto-oncogene whose amplification and overexpression are frequently associated with human cancers (Diehl, 2002). Because its subcellular localization is of critical importance for its oncogenicity, the regulatory mechanisms have been under intense investigation. Here we discovered that the nuclear localization of cyclin D1 was anchorage-dependent, and its disruption caused anchorage-independent growth and survival of cells, a hallmark of cellular transformation. In adherent cells, cyclin D1 was localized in the nucleus by a focal adhesion protein, Hic-5, shuttling in and out of the nucleus through the CRM1 export system (Shibanuma et al. 2003) and thereby counteracting the nuclear export of cyclin D1. In non-adherent cells, cyclin D1 was actively exported from the nucleus because the shuttling of Hic-5, which is redox-sensitive (Shibanuma et al. 2003), was interrupted by an elevated level of reactive oxygen species (ROS). However, when a mutant with the shuttling ability resistant to ROS was introduced into cells, cyclin D1 was detained in the nucleus, and importantly, a significant population of cells survived under non-adherent conditions. Of interest, the discovered phenomenon interconnected the oncogenic potential of two oncogenes, as activated ras circumvented the above regulation and achieved the predominant nuclear localization of cyclin D1 and thus, growth in non-adherent cells.

Key words. cyclin D1, anchorage-dependent growth, Hic-5, ROS

Introduction

The long-standing interpretation of the oncogenecity of cyclin D1 has recently been challenged. Researchers have encountered several cases suggesting that the oncogenic function of cyclin D1 is not simply to promote cell-cycle progression in a cdk-dependent manner (Ewen and Lamb, 2004; Sherr and Roberts, 2004), and intensive study of the cycin D-null mouse failed to support the essential role of cyclin D1 in cell-cycle progression (Kozar et al., 2004). It has been also pointed out that it is the nuclear localization, not mere overexpression, that contributes to tumorigenesis (Gladden and Diehl, 2005).

Results

This study started with the initial observation that cyclin D1 displayed clear nuclear localization only in fully adherent cells among a mixed population of cells with different adhesive states as shown in Figure 1a. In this experiment, the cells were detached and replated one population after another on a coverslip, and after a given period, distinguished according to the criteria based on the formation of focal adhesions, which was monitored according to the incorporation of a focal adhesion protein, Hic-5 (Matsuya et al., 1998), into the structures. In fully adhesive cells (FA+; arrow), immunostaining with an antibody to cyclin D1 observed by confocal fluorescence microscopy revealed a distinct nuclear signal, but in sharp contrast to this, only a vague signal was detected in the nucleus of incompletely adhesive cells (FA-; arrowheads). At 120 min after replating, the clear nuclear signal was restored (data not shown). Of note, when cells were pretreated with an inhibitor of CRM1-dependent nuclear export, leptomycin B (LMB), cyclin D1 stayed in the nucleus even in the FA- cells (data not shown). These optical observations were quantified in Figure 1b and c with two parameters, frequency of the nuclear localization and a ratio of the fluorescence intensity of the nuclear signal to that of the entire cell. The nuclear localization of a transcription factor, Sp1, was evaluated in parallel to validate the whole process. Essentially the same results were obtained with other immortalized and primary cells including human normal diploid fibroblasts (data not shown), and also with tagged cyclin D1 exogenously expressed in cells (Fig. 1d; WT). Another G1 cyclin, cyclin E, showed only marginal anchorage-dependency of its nuclear localization (Fig. 1d). Treatment with cytochalasin D interfering with the formation of actin stress fibers and causing eventual detachment of the cells from the substratum also decreased the nuclear localization of cyclin D1 in 60 min (data not shown), excluding any unintended effect of trypsinization in the above experimental procedure. Biochemical fractionation of cells further supported the above results as when compared with the counterpart in monolayers, 3 h incubation of cells in suspension resulted in a decrease in cyclin D1 in the nuclear fraction with a concomitant increase in cytoplasm (Fig. 1e).

Previously, the nuclear export of cyclin D1 at the onset of the S phase was shown to be dependent on phosphorylation at a threonine, Thr-286, by glycogen synthase kinase 3 beta (GSK-3β), and on CRM1 (Alt et al., 2000; Diehl et al., 1998). The decrease in the nuclear localization of cyclin D1 observed above upon loss of adhesion, which was examined with cell populations at random phases of the cell-cycle, was also dependent on CRM1 (Fig. 1b, c). However, unlike S phase-dependent regulation, the export was dependent on neither the phosphorylation at Thr-286 (Fig. 1d, T286A) nor GSK-3β (data not shown). In line with this, we found that non-phosphorylative T286A mutant was also exported from the nucleus by CRM1 as well as wild-type cyclin D1, albeit with slightly lower efficiency (Fig. 1f).

Western blotting showed that the total amount of cyclin D1 decreased in 4 h incubation in suspension and that the decrease was prevented by inhibition of the proteasome activity by MG132 (data not shown). Given that cyclin D1 was reportedly subjected to proteolysis by the ubiquitin-proteasome pathway in the cytoplasm (Diehl et al., 1997), this observation supported the export of cyclin D1 into the cytoplasm in suspension. On the other hand, this result raised the possibility that the observed decrease in the nuclear localization of cyclin D1 upon loss of adhesion was an artifact of the net decrease in the amount of protein or somehow dependent on proteolysis in the cytoplasm. However, essentially the same decrease in the nuclear localization of cyclin D1 was observed in FA- cells in culture supplemented with MG132 (data not shown).

Based on the above results, we concluded that the nuclear localization of cyclin D1 was anchorage-dependent, and upon loss of adequate adhesion to the substratum, cyclin D1 was transported out of the nucleus by the CRM1 nuclear export system.

Hic-5 is a multidomain LIM protein providing a molecular scaffold for a variety of cellular activities, including integrin signaling at focal adhesions and transcriptional activities in the nucleus (Guerrero-Santoro et al., 2004; Nishiya et al., 2001; Shibanuma et al., 2004). In addition, it shuttles between cytoplasmic and nuclear compartments dependent on CRM1, coordinating the adhesion signal with cellular activities in the two

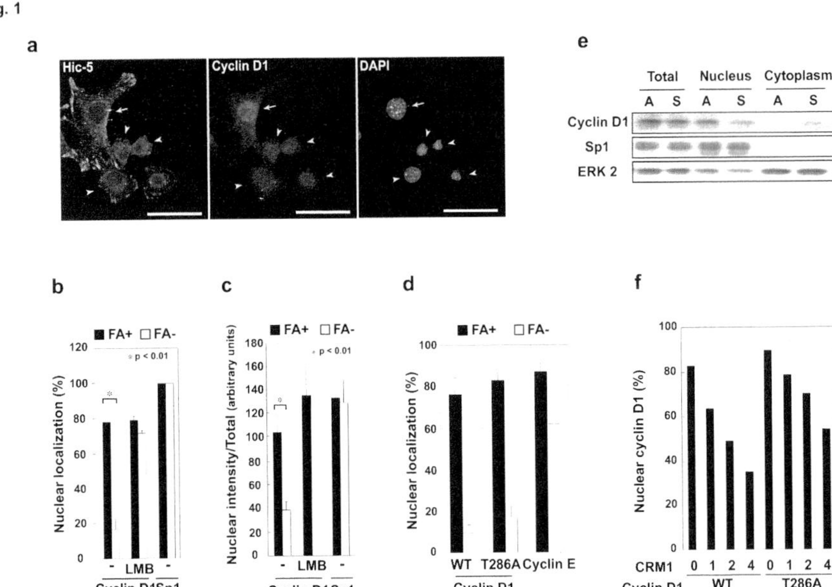

Fig. 1. Anchorage-dependency of the nuclear localization of cyclin D1. (**a**) C3H10T1/2 cells were trypsinized, divided into two populations, and then replated one after another on a coverslip at 90 min intervals, allowing each population to adhere for 120 min or 30 min in total. After fixation, the cells were processed for immunocytochemistry with the antibodies to cyclin D1 and Hic-5. The nuclei were labeled with 4’, 6-diamino-2-phenylindole (DAPI). Photographs were taken under a confocal fluorescence microscope. The arrow and arrowheads indicate FA+ and FAcells, respectively. The scale bars represent 50 μm. (**b**, **c**) The cells pretreated with or without LMB (10 ng/ml) for 1 h were suspended, replated as in **a**, and co-immunostained with the antibody to Hic-5 along with that to cyclin D1 or Sp1. The cells were distinguished as FA+ or FA- according to the criteria, and in **b**, those with dominant nuclear signals, as observed in the FA+ cells in **a** (cyclin D1), for cyclin D1 and Sp1 were optically numerated in each population under a fluorescence microscope, and the ratio is shown. At least 50 cells of each population were scored. In **c**, the fluorescence intensities of the nucleus and the whole of individual cells were measured on the images acquired with the BZ-analyzer, and the ratio of the intensity of the nucleus to that of the whole cell was graphed. In **b** and **c**, the means $\pm$ S.D. obtained from more than three independent experiments are shown. (**d**) The expression vectors for HA-tagged wild-type (WT), T286A mutant of cyclin D1, and cyclin E were introduced into the cells, and after 24 h, the cells were detached and replated as in **a**. The proteins were visualized by co-immunostaining with the antibodies to HA-tag and Hic-5, and the frequency of the nuclear localization of HA-tagged cyclin D1 (WT, T286A) and cyclin E was examined as in **b**. (**e**) NMuMG cells pretreated with MG132 (20 μM) for 1 h were incubated in monolayer (A) or suspension (S)

cultures for 3 h. The cells were subsequently fractionated into nuclear and cytoplasmic fractions and analyzed by Western blotting using the antibodies to cyclin D1, Sp1 and ERK2. Sp1 localized to the nucleus and ERK2 predominantly distributed in the cytoplasm served as monitors of successful fractionation as well as loading controls of the proteins. (**f**) The expression vectors for the HA-tagged wild-type (WT) or T286A mutant cyclin D1 were introduced into C3H10T1/2 cells together with the vector for myc-tagged CRM1 at the indicated CRM1:cyclin D1 ratios (0; 0:1, 1; 1:1, 2; 2:1, 4; 4:1). At 24 h after the introduction, the cells in monolayers were immuostained with the antibody to the HA-tag, and the frequency of the nuclear localization of HA-tagged cyclin D1 was examined as in **b**.

compartments (Mori et al., 2006). It should be emphasized that its shuttling was distinctive in that it was sensitive to cellular redox state and blunted by oxidants (Shibanuma et al., 2003). Here we investigated the involvement of Hic-5 in the adhesion-dependent regulation of the nuclear localization of cyclin D1 using a siRNA against Hic-5 that showed a specific knockdown effect (data not shown) and found that two unrelated siRNA sequences appreciably reduced the nuclear localization of cyclin D1 in monolayers (Fig. 2a). The nuclear localization of both the wild-type and T286A mutant cyclin D1 were equally reduced by the siRNA (Fig. 2b). In Figure 2c, in turn, the effect of overexpression of Hic-5 was examined. Overexpression of Hic-5 but not paxillin, most homologous to Hic-5 (Thomas et al., 1999), increased the nuclear retention of cyclin D1 in FA-cells. Taken together, Hic-5 was suggested to play a role in actively localizing cyclin D1 to the nucleus.

Given that the nuclear export of both Hic-5 and cyclin D1 was dependent on the CRM1 nuclear export system as above, we speculated that Hic-5, while shuttling, competed with cyclin D1 for nuclear export, detaining cyclin D1 in the nucleus. Therefore, we studied the effect of Hic-5 and nuclear export signal (NES)-mutants thereof on the nuclear export of cyclin D1 driven by CRM1 as shown in Figure 2d. While CRM1 decreased the nuclear localization of cyclin D1 in adherent cells (Fig. 2d, CRM1/Hic/PINCH; -/-/- vs +/-/-), co-expression of wild-type Hic-5 counteracted the effect of CRM1 as expected (+/-/- vs +/W/-). The effect of Hic-5 was further augmented by the presence of PINCH, a shuttling partner (Mori et al., 2006) (+/W/- vs +/W/+). Importantly, one of the Hic-5 mutants (LD) whose shuttling ability was disrupted lost completely the counteracting ability (+/W/- vs +/LD/-). Another type of mutant (Cfl/ns) that retained the shuttling capability itself as below but lost specifically the sensitivity of the shuttling to oxidants (Shibanuma et al., 2003) showed even stronger counteracting ability than the wild type (+/W/- vs +/Cfl/ns/-).

This result highlighted the importance of the shuttling of Hic-5 in the effect, supporting the above assumption that Hic-5 competes with cyclin D1 for the CRM1 nuclear export system by its shuttling and as a consequence, promotes the nuclear localization of cyclin D1.

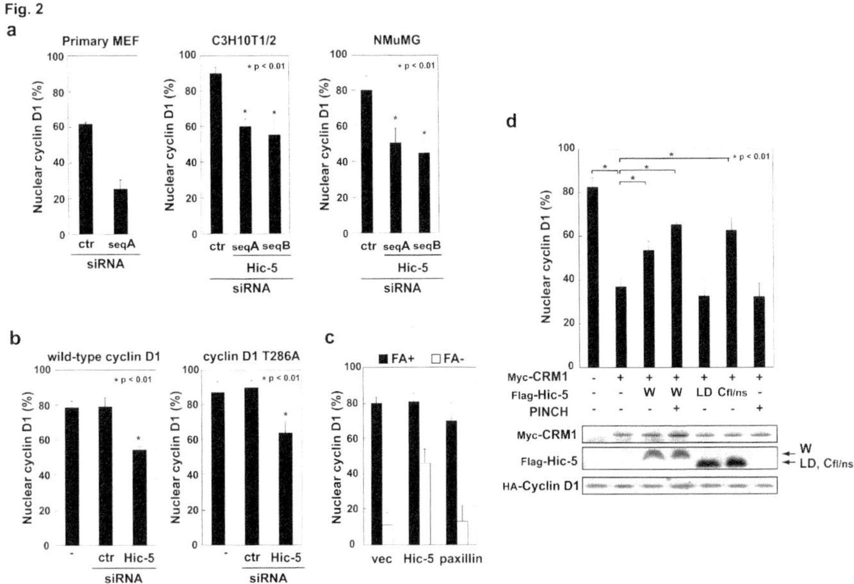

Fig. 2. The role of Hic-5, a focal adhesion protein shuttling between nuclear and cytoplasmic compartments, in the nuclear localization of cyclin D1. (**a**, **b**) The siRNA against Hic-5 (seqA or seqB) or control siRNA (ctr) was introduced into cells (**a**; primary MEF, C3H10T1/2 and NMuMG, **b**; C3H10T1/2). In **b**, the expression vectors for HA-tagged wild-type or T286A mutant cyclin D1 were co-transfected with the siRNA. After 48 h, the cells in monolayers were immunostained with the antibody to cyclin D1 (**a**) or the HA-tag (**b**), and the frequency of the nuclear localization of endogenous (**a**) and exogenous (**b**) cyclin D1 was quantified as in Figure 1b. (**c**) HA-tagged Hic-5 or paxillin was expressed from the expression vector in C3H10T1/2 cells, and at 24 h post-transfection, the cells were detached and replated as in Fig 1a, followed by co-immunostaining with the antibodies to cyclin D1 and the HA-tag. The frequency of the nuclear localization of cyclin D1 in the cells expressing HA-tagged Hic-5 or paxillin was assessed as in Figure 1b. (**d**) The expression vectors for myc-tagged CRM1, Flag-tagged wild-type (W) or mutant Hic-5 (LD; mLD3 or Cfl/ns), and Flag-tagged PINCH were introduced into C3H10T1/2 cells in the indicated combinations together with HA-tagged cyclin D1. The nuclear localization of cyclin D1 was visualized in monolayers by immunostaining with the antibody to HA-tag and, its frequency was assessed as above. The expression levels of each protein in the combinations were examined by Western blotting with the antibodies to each tag and are shown at the bottom.

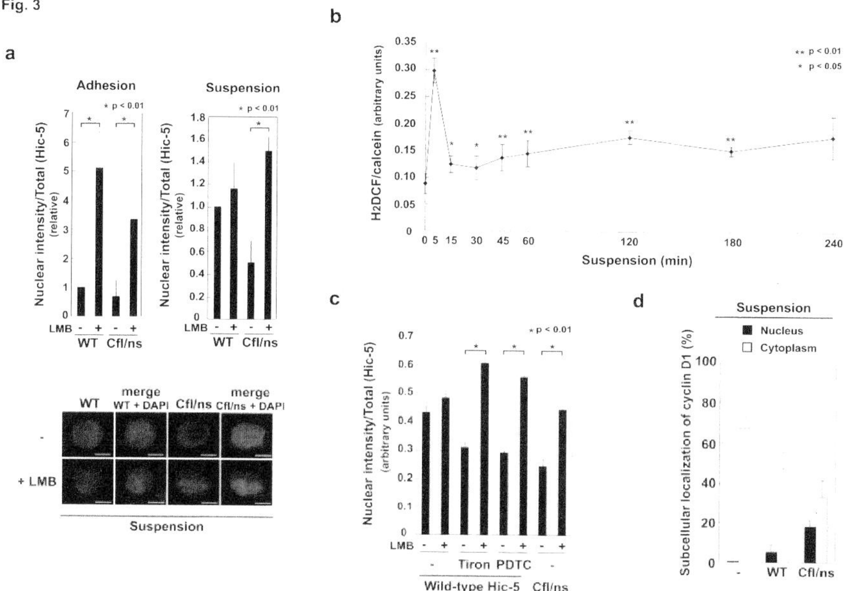

Fig. 3. Anchorage-dependent nuclear-cytoplasmic shuttling of Hic-5 interrupted by elevated levels of ROS in non-adherent cells. (**a**) The vectors for HA-tagged wild-type (WT) or mutant Hic-5, Cfl/ns, were introduced in C3H10T1/2 cells. After 24 h, the cells were detached, transferred to a suspension culture (Suspension) or replated onto coverslips (Adhesion), and further incubated for 2 h with or without LMB (10 ng/ml). After immunostaining with the antibody to the HA-tag, the ratio of the fluorescence intensity of the nucleus to that of the whole cell was obtained and graphed. At least 50 cells of each culture were scored, and the means ±S.D. obtained from more than three independent experiments are shown. Typical images obtained in the suspension culture are presented at the bottom. The nuclei were labeled with DAPI. (**b**) C3H10T1/2 cells were detached, transferred to a suspension culture and after the indicated periods, the level of ROS was measured. In each experiment, more than 20 cells were examined, and the experiment was repeated five times. The values are means ± S.D. (**c**) The cells expressing HA-tagged wild-type Hic-5 as in **a** were pre-incubated with ROS scavengers (1 mM Tiron and 100 μM PDTC) for 30 min, detached and transferred to a suspension culture, and then, incubated 2 h with or without LMB (10 ng/ml). The nuclear localization of Hic-5 was monitored as in **a**. (**d**) The cells expressing Flag-tagged wild-type (WT) or Cfl/ns mutant Hic-5 together with the HA-tagged cyclin D1 were transferred to suspension cultures and after 2 h, immunostained with the antibody to the HA-tag. The numbers of cells with cyclin D1 dominantly distributed in the nucleus and in the cytoplasm were numerated and are shown as a ratio. Filled and open bars indicate the predominant nuclear and cytoplasmic localization of cyclin D1, respectively. The means ± S.D. obtained from independent experiments repeated more than three times are shown. Refer to color plates.

Based on the above findings, it was most likely that in non-adherent cells, Hic-5 stopped the shuttling, resulting in increased availability of the nuclear export system for cyclin D1 and thus, leading to its transport out of the nucleus. The shuttling of Hic-5, evaluated by LMB-sensitive nuclear accumulation (Shibanuma et al., 2003), was in fact found to be severely impeded in non-adherent cells (Fig. 3a, Suspension; WT). Interestingly, the Hic-5 mutant, Cfl/ns, retained the capability and shuttled in non-adherent cells as in adherent ones (Suspension; Cfl/ns). Considering that Cfl/ns was the mutant that was manipulated to acquire the shuttling ability resistant to the inactivation by oxidants through the mutation at specific cysteine residues (Shibanuma et al., 2003), it was predictable that cellular redox state in non-adherent cells shifted to oxidative conditions so that the shuttling of the wild type was blunted. To prove this assumption, we first monitored the level of ROS in the cells detached from the substratum by using a fluorescence indicator and observed a burst of ROS production followed by a sustained increase (Fig. 3b). Subsequently, we tested the effect of scavenging ROS and observed that when the cells were pretreated with chemical scavengers of ROS, 1,2-dihidroxybenzene-3,5-disulphonic acid (Tiron) and pyrrolidine dithiocarbamate (PDTC), even wild-type Hic-5 retained the shuttling capability in non-adherent cells (Fig. 3c). These observations provided substantial evidence that under non-adherent conditions, the increased production of ROS halted the shuttling of Hic-5. Then, the C/fl/ns mutant that is able to shuttle independently of oxidative conditions hypothetically localizes cyclin D1 to the nucleus even in non-adherent cells. This was the case, and the introduction of the mutant in the cells increased the population of the cells in which cyclin D1 was localized to the nucleus in the suspended culture (Fig. 3d). Besides, this result argued that the shuttling of Hic-5 indeed acted as a driving force to localize cyclin D1 to the nucleus. Next we addressed the significance of the anchorage-dependency of the nuclear localization of cyclin D1 regulated by the redox-sensitive shuttling of Hic-5 and a consequence of its deregulation resulting in the nuclear localization of cyclin D1 in non-adherent cells. For this purpose, we retrovirally overexpressed the wild-type and oxidant-resistant Cfl/ns mutant of Hic-5, or NLS-cyclinD1, a derivative of cyclin D1 with a nuclear localization signal (NLS), in mouse mammary epithelial NMuMG cells, placed the cells in suspension and performed BrdU incorporation analysis. As shown in Figure 4a, overexpression of NLS-cyclin D1 prominently stimulated BrdU incorporation 4-fold (D1/Hic-5; NLS/-). Of note, the C/fl/ns mutant also increased the incorporation more than 2-fold over the control by itself (D1/Hic-5; -/Cfl/ns). Like in NMuMG cells, in the suspension culture of

C3H10T1/2 cells, the amount of BrdU incorporated was increased by the introduction of NLS-cyclin D1 (Fig. 4b, D1/Hic-5; NLS/v). In comparison, the effect of wild-type cyclin D1 was much weaker in accordance with previous reports (Quelle et al., 1993; Resnitzky et al., 1994) (D1/Hic-5; +/v), and the C/fl/ns mutant, when co-expressed with wild-type cyclin D1, potentiated its effect (D1/Hic-5; +/Cfl/ns). In parallel, modest but reproducible increases in the fraction of S phase were observed by the flow cytometry (data not shown). In the suspension culture of NMuMG cells, we noticed the appearance of a population with sub G1 DNA content using flow cytometry (data not shown), suggesting that a fraction of the epithelial cells underwent apoptosis. In fact, the apoptotic cells were detected in the suspension culture, and the population was reduced by the expression of the Cfl/ns mutant as well as NLS-cyclin D1 (Fig. 4c), suggesting that the nuclear-localized cyclin D1 had an anti-apoptotic effect as well as the growth promoting activity in the suspended cells. Consequently and importantly, significant numbers of cells were alive after the long incubation in suspension in the presence of Cfl/ns of Hic-5 and NLS-cyclin D1 (Fig. 4d). Overall, these analyses demonstrated a partial but appreciable escape from growth arrest and/or acquisition of resistance to apoptosis in the non-adherent cells by ROS-insensitive continuous shuttling of Hic-5, increasing the nuclear localization of cyclin D1.

Finally, to implicate the above phenomenon in tumorigenesis, we introduced v-Ki-*ras* into cells and examined the nuclear localization of cyclin D. Of particular interest, in the cells expressing v-Ki-*ras,* intense nuclear signals were observed in the suspension culture rather than in the monolayer (Fig. 5a and b). The nuclear signal was diminished by knockdown of cyclin D1 with siRNA, ruling out the cross-reactivity of the antibody to an irrelevant protein (data not shown). Most importantly, in the presence of the siRNA against cyclin D1, the cells expressing *ras* failed to incorporate BrdU any more than control cells in suspension culture (Fig. 5c), suggesting that *ras* was dependent on the nuclear localization of cyclin D1 to induce anchorage-independent growth of cells (Yang et al., 1998). This result also provided a mechanistic insight into the requirement of cyclin D1 in tumorigenesis by *ras* reported previously (Yu et al., 2001). In addition, the signal of *ras* was likely to modify cyclin D1 itself to escape from the regulatory mechanism provided by Hic-5 rather than to modify the shuttling of Hic-5, since the shuttling was interrupted in the suspended cells expressing *ras* as well as non-expressing cells (Fig. 5d).

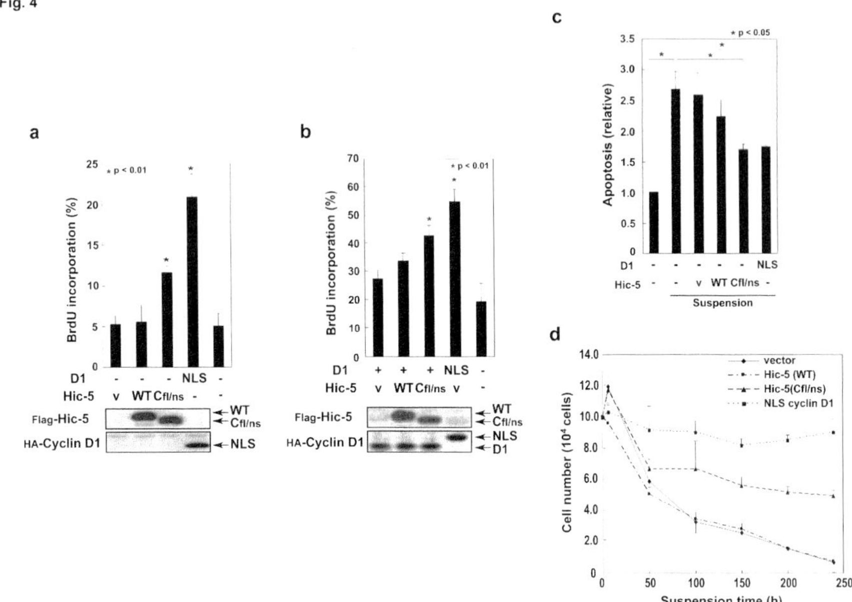

Fig. 4. Escape from growth arrest and acquisition of resistance to apoptosis induced by oxidant-resistant shuttling of Hic-5 and nuclear-targeted cyclin D1 in suspension culture. (**a**) NMuMG cells were infected with one of the following retroviral vectors; the empty vector (v), the vector encoding Flag-tagged wild-type (WT) or Cfl/ns mutant Hic-5, and the vector encoding HA-tagged nuclear-targeted cyclin D1 (NLS). At 72 h post-infection, the cells were placed in suspension for 24 h, and subsequently, incorporation of BrdU was examined after another 48 h. The experiments were repeated more than three times using independent sets of infected samples. The means ± S.D. are shown. In the lower panel, total lysates of the cells infected as above were subjected to Western blotting with the antibodies to the tags. (**b**) C3H10T1/2 cells were sequentially infected with retroviral vectors, first with an empty vector (v) or those encoding Flag-tagged wild-type (WT) or Cfl/ns mutant Hic-5 and then, with those encoding HA-tagged wild-type (D1) or nuclear-targeted cyclin D1 (NLS). At 72 h after the second infection, the cells were placed in suspension for 24 h, and the incorporation of BrdU was examined after another 12 h. In the lower panel, total lysates of the cells infected as above were subjected to Western blotting with the antibodies to the tags. (**c**) The NMuMG cells infected with the indicated retroviral vectors as in **a** and placed in suspension for 48 h were examined with a colorimetric "APOPercentage" apoptosis assay kit according to the directions provided. The population of the apoptotic cells was quantified by releasing the incorporated dye from the cells and measuring the absorbance at 570 nm. The values were normalized with the total number of cells, and those relative to the adherent condition are shown as means ± S.D. from more than three independent assays. (**d**) The NMuMG cells infected as in **a** were transferred to suspension cultures, and the number of cells was counted at the indicated time.

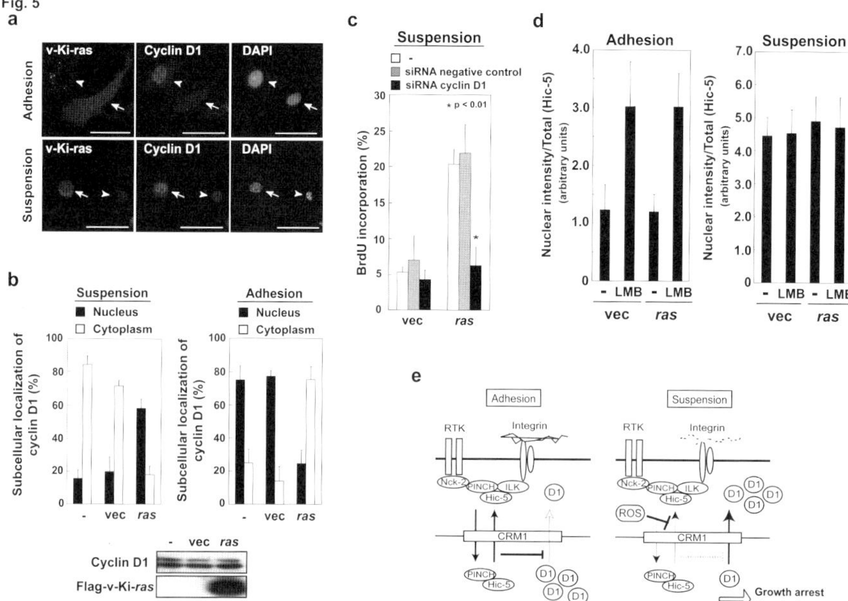

Fig. 5. Disruption of the anchorage dependency of the nuclear localization of cyclin D1 by oncogenic ras. (**a**) The expression vector encoding Flag-tagged v-Ki-ras was introduced into NIH3T3 cells, and 24 h later, the cells were detached, and transferred to suspension culture (Suspension) or cultured in monolayers (Adhesion). After 3 h of incubation, the cells were fixed and co-immunostained with the antibodies to cyclin D1 and the Flag-tag. The nuclei were labeled with DAPI. Photographs were taken under a confocal fluorescence microscope. The arrow and arrowhead indicate cells expressing (Flag-positive) and not expressing (Flag-negative) ras, respectively. The scale bars represent 50 μm. (**b**) The expression vector encoding the Flag-tagged v-Ki-ras (*ras*) or the empty vector (vec) was introduced into the cells, and processed as in **a**. Subsequently, the subcellular localization of cyclin D1 was examined in the Flag-positive cells as in Figure 3d. Total cell lysates were analyzed by Western blotting with the antibodies to cyclin D1 and the Flag-tag at the bottom. (**c**) The expression vector for the ras or empty vector was introduced into the cells with siRNA for cyclin D1 or control and 48 h later, the cells were placed in suspension and further incubated for 24 h. Then, BrdU was added and after 4 h of incubation, the incorporation was examined as in Fig. 4a,b in the cells after co-immunostaining with the antibodies to the tag and BrdU. (**d**) The vectors for HA-tagged wild-type Hic-5 and the Flag-tagged v-Ki-ras were introduced into C3H10T1/2 cells. After 24 h, the cells were detached, transferred to suspension cultures (Suspension) or replated onto coverslips (Adhesion), and further incubated for 2 h with or without LMB (10 ng/ml). The cells were immunostained with the antibodies to the tags, and the nuclear localization of Hic-5 in the cells was assessed as in Fig. 3a. (**e**) A model of a failsafe system for anchorage-dependence of cell growth and survival. Cyclin D1 was localized to the nucleus in adherent cells by the nuclear-cytoplasmic shuttling of Hic-5, which counteracted the nuclear export of cyclin D1. In suspended cells, the shuttling was blunted by the elevated levels of ROS and thus, cyclin D1 was actively transported out of the nucleus, resulting in a decrease in its nuclear localizaiton. Otherwise, the cells acquire the capability to survive and/or grow in anchorage-independent conditions.

Conclusion

Altogether, we summarized the findings of this study in a model proposing a failsafe system for anchorage-dependence of cell growth and survival, the system coupling the cellular growth and survival strictly with an appropriate matrix attachment through the nuclear localization of cyclin D1 (Fig. 5e). Otherwise, irrelevant cell-cycle progression or resistance to apoptosis in non-adherent cells most likely contributes to tumorigenesis, increasing the survival of cells in invading and circulating processes. The mechanisms underlying the escape from growth arrest or apoptosis by the nuclear-localized cyclin D1 deserve further investigation. The phosphorylation of Rb at serine 780 by cyclin D1-cdk4/6 showed no robust changes under the above conditions (data not shown), making the involvement of hyperphosphorylation of Rb, a classical hallmark of the transition from G1 to S phase, obscure. An increasing number of studies have posed a new framework viewing cyclin D1 as a transcriptional coregulator (Coqueret, 2002; Lamb et al., 2003).

According to recent progress, cyclin D was not required for proliferation in most types of cells *in vivo* except hematopoietic stem cells (Kozar et al., 2004; Sherr and Roberts, 2004), suggesting the possible difference of the functionality of cyclin D1 dependent on adhesive conditions and a pro-proliferative potential of the cyclin D1 distinctively manifested in non-adherent cells. This possibility implied the requirement of a failsafe system as above to prevent irrelevant availability of the pro-proliferative potential in case of detachment or inappropriate matrix attachment of adherent cells. Conversely, as was the case for *ras*, molecular mechanisms underlying the anchorage-dependency of the nuclear localization of cyclin D1 could be a reasonable target of oncogenic signals. This study will hopefully contribute to the development of new therapies for malignancies designed to deprive the target protein of oncogenic potential through manipulation of subcellular localization, keeping the protein away from the sites of action.

References

Alt, J.R., J.L. Cleveland, M. Hannink, and J.A. Diehl. (2000) Phosphorylation-dependent regulation of cyclin D1 nuclear export and cyclin D1-dependent cellular transformation. Genes Dev. 14:3102-14.

Coqueret, O. (2002) Linking cyclins to transcriptional control. Gene. 299:35-55.

Diehl, J.A. (2002) Cycling to cancer with cyclin D1. Cancer Biol Ther. 1:226-31.

Diehl, J.A., M. Cheng, M.F. Roussel, and C.J. Sherr. (1998) Glycogen synthase kinase-3beta regulates cyclin D1 proteolysis and subcellular localization. Genes Dev. 12:3499-511.

Diehl, J.A., F. Zindy, and C.J. Sherr. (1997) Inhibition of cyclin D1 phosphorylation on threonine-286 prevents its rapid degradation via the ubiquitin-proteasome pathway. Genes Dev. 11:957-72.

Ewen, M.E., and J. Lamb. (2004) The activities of cyclin D1 that drive tumorigenesis. Trends Mol Med. 10:158-62.

Gladden, A.B., and J.A. Diehl. (2005) Location, location, location: the role of cyclin D1 nuclear localization in cancer. J Cell Biochem. 96:906-13.

Guerrero-Santoro, J., L. Yang, M.R. Stallcup, and D.B. DeFranco. (2004) Distinct LIM domains of Hic-5/ARA55 are required for nuclear matrix targeting and glucocorticoid receptor binding and coactivation. J Cell Biochem. 92:810-9.

Kozar, K., M.A. Ciemerych, V.I. Rebel, H. Shigematsu, A. Zagozdzon, E. Sicinska, Y. Geng, Q. Yu, S. Bhattacharya, R.T. Bronson, K. Akashi, and P. Sicinski. (2004) Mouse development and cell proliferation in the absence of D-cyclins. Cell. 118:477-91.

Lamb, J., S. Ramaswamy, H.L. Ford, B. Contreras, R.V. Martinez, F.S. Kittrell, C.A. Zahnow, N. Patterson, T.R. Golub, and M.E. Ewen. (2003) A mechanism of cyclin D1 action encoded in the patterns of gene expression in human cancer. Cell. 114:323-34.

Matsuya, M., H. Sasaki, H. Aoto, T. Mitaka, K. Nagura, T. Ohba, M. Ishino, S. Takahashi, R. Suzuki, and T. Sasaki. (1998) Cell adhesion kinase beta forms a complex with a new member, Hic-5, of proteins localized at focal adhesions. J Biol Chem. 273:1003-14.

Mori, K., M. Asakawa, M. Hayashi, M. Imura, T. Ohki, E. Hirao, J.R. Kim-Kaneyama, K. Nose, and M. Shibanuma. (2006) Oligomerizing potential of a focal adhesion LIM protein Hic-5 organizing a nuclear-cytoplasmic shuttling complex. J Biol Chem. 281:22048-61.

Nishiya, N., K. Tachibana, M. Shibanuma, J.I. Mashimo, and K. Nose. (2001) Hic-5-reduced cell spreading on fibronectin: competitive effects between paxillin and Hic-5 through interaction with focal adhesion kinase. Mol Cell Biol. 21:5332-45.

Quelle, D.E., R.A. Ashmun, S.A. Shurtleff, J.Y. Kato, D. Bar-Sagi, M.F. Roussel, and C.J. Sherr. (1993) Overexpression of mouse D-type cyclins accelerates G1 phase in rodent fibroblasts. Genes Dev. 7:1559-71.

Resnitzky, D., M. Gossen, H. Bujard, and S.I. Reed. (1994) Acceleration of the G1/S phase transition by expression of cyclins D1 and E with an inducible system. Mol Cell Biol. 14:1669-79.

Sherr, C.J., and J.M. Roberts. (2004) Living with or without cyclins and cyclin-dependent kinases. Genes Dev. 18:2699-711.

Shibanuma, M., J.R. Kim-Kaneyama, K. Ishino, N. Sakamoto, T. Hishiki, K. Yamaguchi, K. Mori, J. Mashimo, and K. Nose. (2003) Hic-5 communicates between focal adhesions and the nucleus through oxidant-sensitive nuclear export signal. Mol Biol Cell. 14:1158-71.

Shibanuma, M., J.R. Kim-Kaneyama, S. Sato, and K. Nose. (2004) A LIM protein, Hic-5, functions as a potential coactivator for Sp1. J Cell Biochem. 91:633-45.

Thomas, S.M., M. Hagel, and C.E. Turner. (1999) Characterization of a focal adhesion protein, Hic-5, that shares extensive homology with paxillin. J Cell Sci. 112 (Pt 2):181-90.

Yang, J.J., J.S. Kang, and R.S. Krauss. (1998) Ras signals to the cell cycle machinery via multiple pathways to induce anchorage-independent growth. Mol Cell Biol. 18:2586-95.

Yu, Q., Y. Geng, and P. Sicinski. (2001) Specific protection against breast cancers by cyclin D1 ablation. Nature. 411:1017-21.

Induction of Matrix Metalloprotease Gene Expression in an Endothelial Cell Line by Direct Interaction with Malignant Cells

Kiyoshi Egawa, Yuki Hasebe, Motoko Shibanuma and Kiyoshi Nose

Department of Microbiology, Showa University School of Pharmaceutical Sciences, Hatanodai 1-5-8, Shinagawa-ku, Tokyo 142-8555, Japan

Summary. Mouse endothelial TKD2 cells in the monolayer were co-cultured with various human cell lines for 24 hr, and the expressions of several secreted MMPs and cell adhesion molecules were examined by real-time RT-PCR using mouse-specific primers. Co-culture with normal fibroblasts did not elicit the expression of these molecules, but co-culture with cancer cells induced the expression of MMP-3, -9, and -10 mRNA in endothelial cells and in normal mouse embryonic fibroblasts. The induction of MMP mRNAs was dependent on direct cell adhesion since a separate culture of A549 cells in Boyden chambers did not induce MMP mRNAs, and a neutralizing antibody against VLA-4 abolished the induction. Intracellular ROS levels increased in TKD2 cells following adhesion to cancer cells. ROS scavengers decreased the levels of MMP induction, and roterone, an inhibitor of mitochondrial complex I, strongly suppressed the induction of MMP-3、-9、and -10. These results indicate that the adhesion of cancer cells to endothelial cells activates several distinct signaling pathways to induce the MMP gene expression.

Key words. cell-cell interaction, MMP-9 induction, reactive oxygen species

Results and Discussion

1 Induction of adhesion-related genes in mouse TKD2 cells co-cultured with human cancer cells

Human cancer cells were added to a monolayer of TKD2 cells at a ratio of 1:1, and cultured for 24 hr at 33° C. Total RNA was extracted, and levels of MMP mRNA in TKD2 cells were examined by real-time RT-PCR. The primers did not detect human MMPs. The results shown in Fig. 1 indicate that co-culture with normal human fibroblasts did not affect the levels of MMP mRNAs significantly, but cultivation with cancer cells induced the expressions of MMP-3, MMP-9 and MMP-10. Levels of MMP-2 also increased but to a lesser extent. Expressions of several cell adhesion molecules such as VCAM-1 and ICAM-1 was also increased in TKD2 cells co-cultured with tumor cells. The effect of A549 seemed most prominent, so we focused on A549 in later experiments.

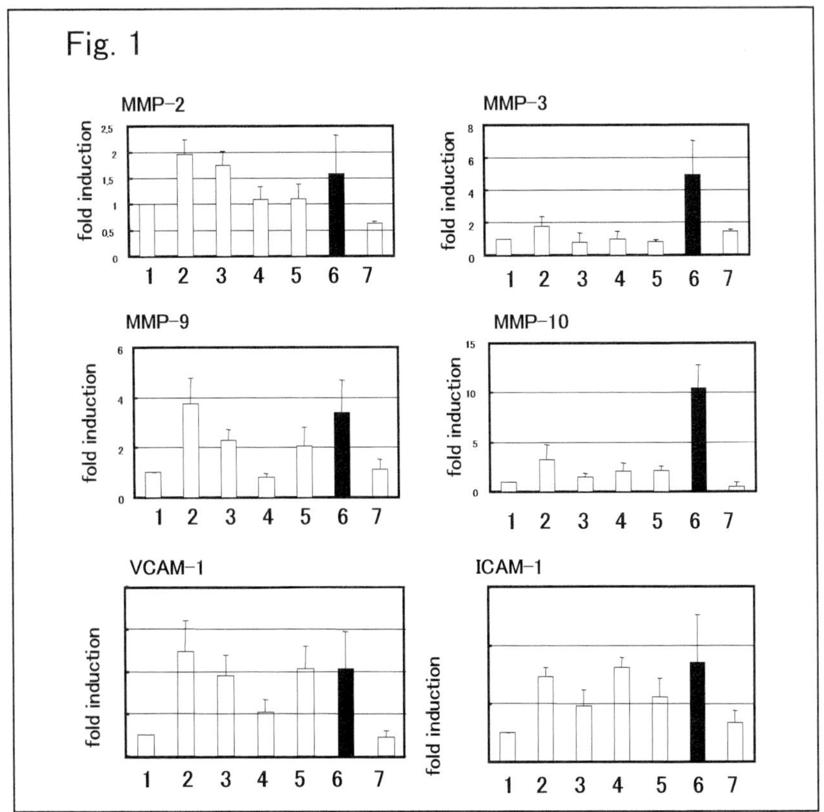

Fig. 1. Changes in MMP-mRNAs in TKD2 cells cultured with human cells. 1, TKD2 alone; 2, TKD2+HT1080; 3, TKD2+TMK-1; 4, TKD2+HepG2; 5, TKD2+HelaS3; 6, TKD2+A549; 7, TKD2+TIG3. Levels of mRNA were measured by quantitative RT-PCR.

A549 cells were seeded on a monolayer of TKD2 cells, and RNA was extracted at various times after starting the co-culture. The MMP-9 mRNA level increased rapidly and reached a plateau at 1 hr. Induction of MMP-3 and MMP-10 was slower, starting after 4 hr and reaching a plateau around 24 to 48 hr.

A549 cells were seeded on a monolayer of TKD2 cells, and anti-integrin α2 antibody or anti-integrin α4 antibody (10μg/mL each) was added at the same time. MMP mRNA levels were quantitated 24 hr after starting the co-culture. The main ligand of integrin α2 is collagen, and that of integrin α4 is VCAM-1. Anti-integrinα4 inhibited the induction of MMP-3, MMP-9 and MMP-10, whereas anti-integrin α2 inhibited the induction of only MMP-9.

2 Signals from adhesion to MMP induction

After the adhesion of TKD2 cells to A549 cells, levels of intracellular reactive oxygen species (ROS) increased as determined by a fluorescent redox sensor, Red CC-1 The importance of ROS in MMP induction was estimated by examining the effects of ROS scavengers and inhibitors of ROS production. An anti-oxidant that scavenges hydroxyradicals, dimethyl thiourea (DMTU), inhibited the induction of MMP-3, -9, and -10. A non-specific inhibitor of flavin-containing enzymes, diphenyleneiodonium(DPI), and an inhibitor of mitochondrial electron transporter complex I, rotenone, strongly inhibited MMP induction (Fig. 2).

Mitochondrial activity seemed to be stimulated in TKD2 cells co-cultured with A549 cells, and this stimulation was inhibited by LY294002 or the dominant-negative form of PI3K, indicating that PI3K activation following cell-cell interaction was required for the activation of mitochondria. MMP-9 induction was significantly decreased in pseudo ρ0 cells and recovered at least in part in revertants. The findings in this study indicate that mitochondria are an important target for the development of anti-metastatic drugs, and further studies are necessary to clearly understand the precise mechanisms.

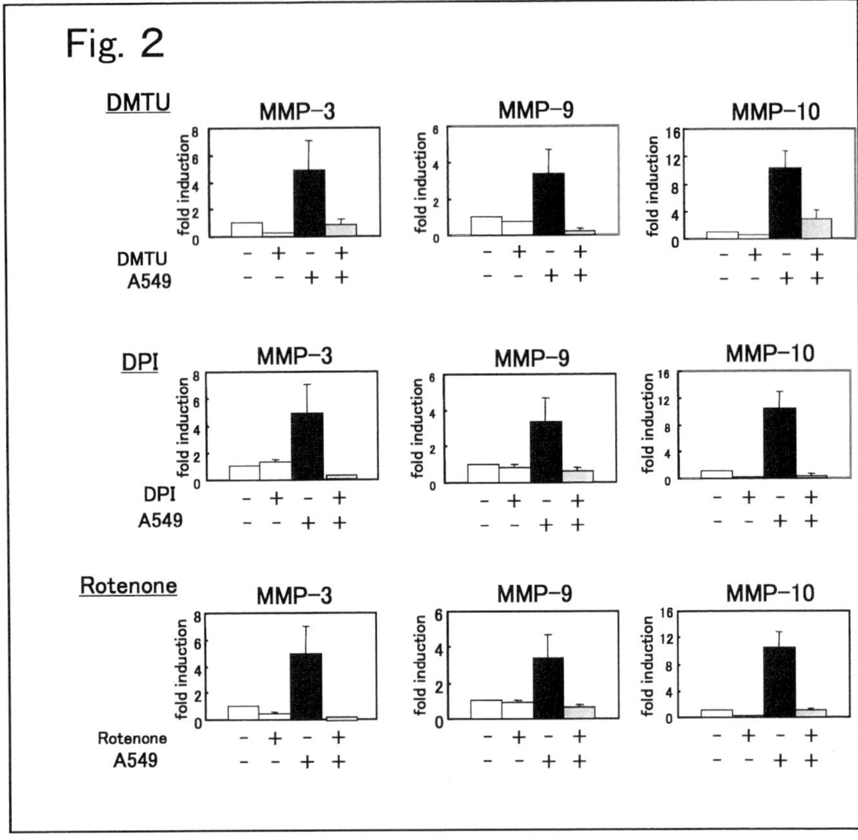

Fig. 2. Effects of radical scavengers on induction of MMP-3, MMP-9 and MMP-10. TKD2 and A549 cells were co-cultured, and were treated with dimethyl thiourea (DMTU), diphenyleneiodonium(DPI), or rotenone. Levels of mRNA were measured by quantitative RT-PCR.

References

Hasebe Y, Egawa K, Shibanuma M, Nose K. (2007) Induction of matrix metalloprotease gene expression in an endothelial cell line by direct interaction with malignant cells. Cancer Sci 98: 58-67

Expression of Rad21 Cleaved Products in Oral Epithelium

Gou Yamamoto, Taku Matsunaga, Tomohide Isobe, Tarou Irie, Tetsuhiko Tachikawa

Department of Oral Pathology and Diagnosis, Showa University School of Dentistry, Hatanodai 1-5-8, Shinagawa-ku, Tokyo 142-8555, Japan

Summary. Although RAD21 is involved in the repair of DNA double-strand breaks and is essential for mitotic growth, its role in cancer has been unclear. Recently, cleaved products of RAD21 by caspase 3 and 7 become a factor inducing apoptosis. And there are some reports about the association of tumor formation and RAD21. Our current study indicates the RAD21 gene was closely related to the invasion and metastasis of the cancer cells. In addition, although it is known that there are many cleavage products, about those functions, it is not clear. In this study, the relevance of RAD21 cleaved products expression to cell differentiation and carcinogenesis of oral squamous epithelium was clarified using Laser Microdissection and Western Blot analysis.

Key words. RAD21, squamous cell carcinoma, laser microdissection

Introduction

RAD21 is one of the four subunits of cohesin, the latter which holds sister chromatides together until anaphase (Birkenbihl and Subramani 1992, Michaelis et al. 1997). Cohesin is combined with the chromosome in the G1 phase and holds the sister chromatides until RAD21 is cleaved by separase during the onset of anaphase, with these cleavages allowing chromosomal separation (Birkenbihl and Subramani 1995, Hirano 2000, Hoque and Ishikawa 2001, Nasmyth et al. 2000, Uhlmann et al. 1999). When RAD21 was knocked out of chicken DT40 cells, it caused premature sister chromatoid separation (Sonoda et al. 2001). When RAD21 is mutated at the cleavage site by separase, not only chromosome segregation but also cell division is inhibited (Hauf et al. 2001). In light of the above fact, RAD21 may play an important role in faithful and stabilized chromosomal separation. Recently, it has been shown that the degradation product formed by the C-terminal cleavage of RAD21 by caspases 3 and 7 becomes a factor which induces apoptosis (Chen et al. 2002, Pati et al. 2002). In addition, in regard to the association of tumor formation and RAD21, the expression level of the RAD21 gene is suppressed by hypoxia in liver cancer and mammary cancer cells (Sook Kim et al. 2001), the RAD21 gene is over expressed in prostate cancer cells (Porkka et al. 2004), and it has been reported that RAD21 gene expression is induced by the BRCA1 gene which is a tumor suppressor gene in breast cancer cells (Atalay et al. 2002). However, the relation between the RAD21 gene and cancer remains unclear. In the present study, the relevance of the RAD21 cleavage products to the invasion and metastasis of squamous cell carcinoma and to the differentiation of oral squamous epithelium were investigated in frozen tissue samples using laser microdissection and western blot method.

Material and Method

Cell culture

The SAS cell, a poorly differentiated squamous cell carcinoma cell line, was derived from human tongue primary lesion (Takahashi et al. 1989). Cells were cultured in Dulbecco's Modified Eagle medium nutrient mixture F-12 ham (SIGMA) supplemented with 10% fetal bovine serum, 50units/ml penicillin, and 50μg/ml streptomycin, in a humidified atmosphere, 5% CO_2, at 37℃. The cell was treated with 5-azacytidine and ALLN at the desired final concentration and duration.

Frozen sample preparation for Laser Microdissection

We used PALM® Micro Beam (P.A.L.M. Microlaser Technologies) for retrieving target area from frozen section. Samples were obtained at surgery. The extracted tissues were embedded in OCT compound and frozen in isopentane cooled in liquid nitrogen. The specimens were then made into frozen blocks, which were sliced using a cryomicrotome (MICROM) at 7μm thicknesses for microdissection.

Western blot analysis

Proteins were extracted from culture cells and frozen sections by RIPA buffer. Cell fractionation was carried out by PARIS kit (AMBION). Western blot analysis was performed according to ordinary method. Anti-RAD21 antibody (BETHYL) was used for primary antibody. For detection of Chemiluminescence, we used Light Capture (ATTO corporation Japan), and analyzed the band density by CS analyzer (ATTO corporation Japan).

Results

1. Our current study

Current study about the relation between RAD21 gene expression and invasion patterns of oral squamou cell carcinoma was shown. The figure indicates microdissected tissue according to each invasion pattern (Fig.1).

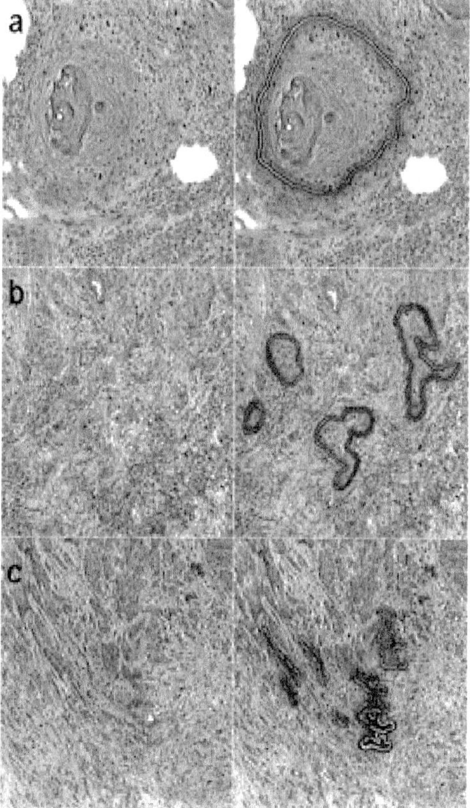

Fig.1 Procurement of the target tissue from the frozen section by the Laser microdissection method. (a) INFα, (b) INFβ, (c) INFγ. Laser microdissection was performed by PALM® Micro Beam (P.A.L.M. Microlaser Technologies) after staining in the 1% toluidine blue. For left-hand side, before the cut by laser and the right are after a cut.

The invasion pattern was determined in terms of the classification (pattern of infiltration; INF) of the cancer handling convention of Japan. The patterns is regarded as INFα if the cancer nest grows expansively and a clear circumference boundary can be seen between the cancer and the surrounding tissue. INFγ shows an infiltrative growth pattern and the boundary with the circumference tissue is indistinct. INFβ is intermediate between INF-alpha and INFγ. As a result, the expression level of the RAD21 gene decreased significantly ($p<0.01$) in the areas of INFβ and INFγ in comparison with INFα. In regional lymph node metastasis , the expression level of RAD21 gene also decreased similar to INFγ(Fig.2).

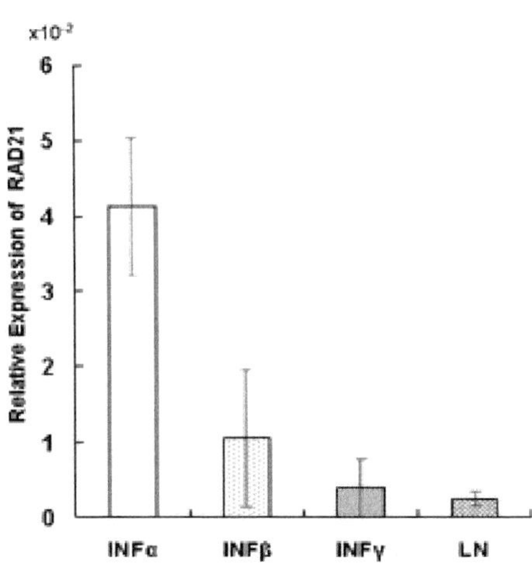

Fig.2 The expression level of the RAD 21 gene by Laser microdissection and Real-Time Quantitative PCR. The relative expression of RAD21 gene was calculated by diving the value by the GAPDH value of each sample. At the infiltrating edge of the oral squamous cell carcinoma, the RAD21 gene expression was decreased in the area that showed the invasion patterns of INFβ (n=8) and INFγ (n=10) in comparision with the area of INFα (n=7). In regional lymph node metastasis (LN: n=5), the RAD21 gene expression was also decreased similar to the INFγ invasion pattern. Bars,±SD.

2. Laser microdissection and western blot

In analyzing expression of RAD21 protein, many cleavage products of RAD21 become a problem. Immunohistochemistry can stain all cleaved protein similarly. But, if we use Laser Microdissection and western blot methods, cleaved protein can be identified according to molecular weight. So we try to detect RAD21 by the methods of Laser Microdissection and western blot. When the methods of Laser Microdissection and western blot was tested, we tried the detection of β-actin first (Fig.3). Though only a very slight band had been detected from samples of about 105 cells, the number of cells was roughly proportional to the expression of β-actin. Y-axis indicates the density of the band analyzed by CS analyzer, X-axis the number of cells.

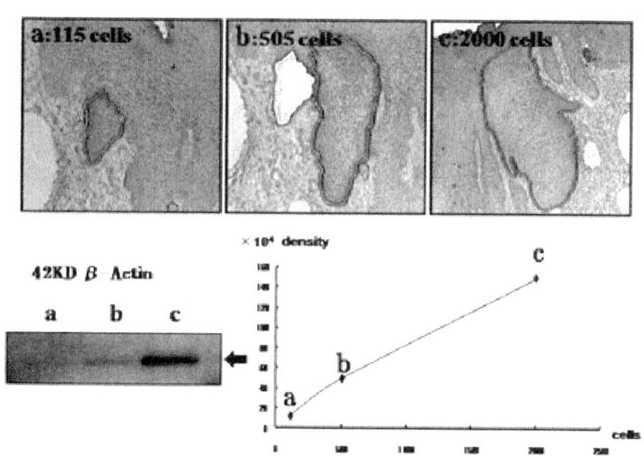

Fig.3 Measurement value of β-Actin using Laser Microdissection and Western Blot method. Indicated number of cells were collected by Laser Microdissection. the number of cells was roughly proportional to the expression of β-actin. Y-axis indicates the density of the band analyzed by CS analyzer, X-axis the number of cells.

3. The expression of RAD21 cleaved products in frozen section

So detecting the protein by Laser Microdissection and western blot became possible, we analyzed RAD21 cleaved products in each invasion pattern (Fig.4). 130KD product is known as full length RAD21. The expression of 80KD and 68KD products has increased in INFα in association with the gene expression.

Because oral squamous cell carcinoma classified into INFα often form large and well differentiated nest, we studied the relation between RAD21 cleaved products and the differentiation of oral squamous epithelium(Fig5). We could detect only a slight band in normal epithelium. But 80KD cleaved product increased in above prickle cell layer of keratinized epithelial hyperplasia and epithelial dysplasia. Especially, in above upper prickle cell layer of epithelial dysplasia, 80KD leaved product was expressed more strongly (Fig.6).

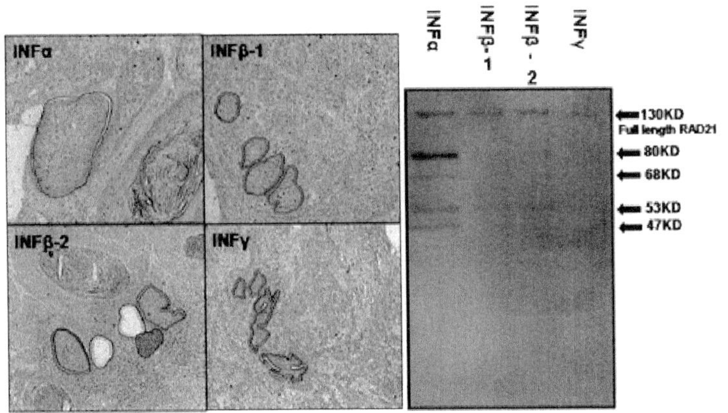

Fig.4 Expression of RAD21 cleaved products in invasion pattern of oral squamous epithelium. 130KD product is known as full length RAD21. The expression of 80KD and 68KD products has increased in INFα in association with the gene expression.

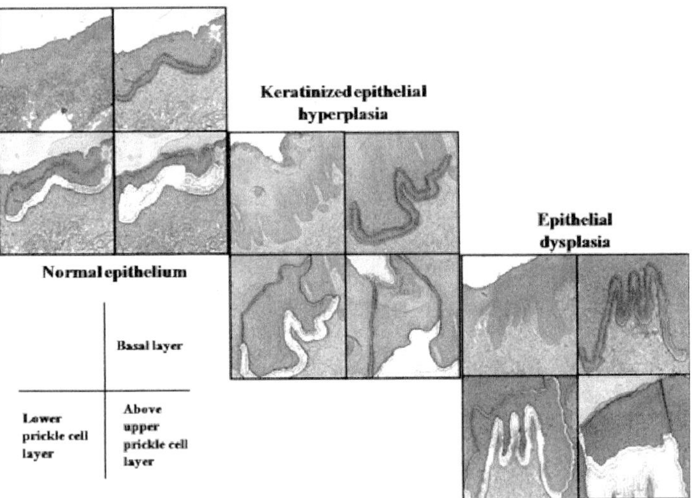

Fig.5 The layers of oral squamous epithelium. Figures indicate each layer; basal layer, lower prickle cell layer and above upper prickle cell layer microdissected from normal epithelium, keratinized epithelial hyperplasia, and epithelial dysplasia. Refer to color plates.

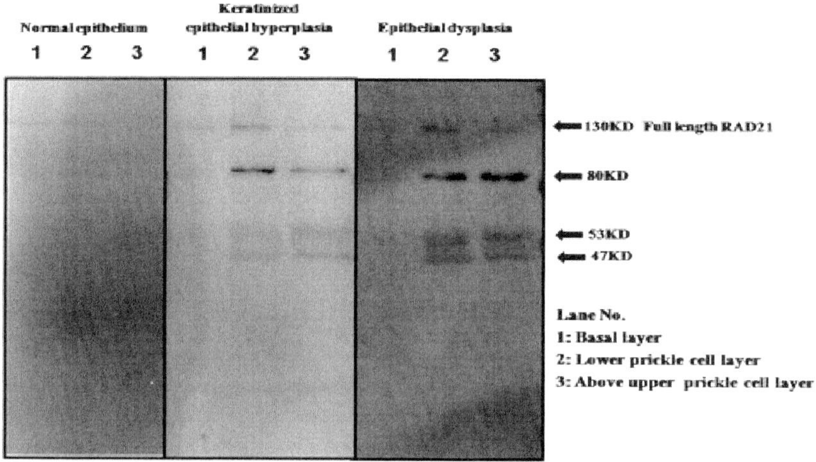

Fig.6. The relation between RAD21 cleaved products and the differentiation of oral squamous epithelium. 80KD cleaved product increased in above prickle cell layer of keratinized epithelial hyperplasia and epithelial dysplasia. Especially, in above upper prickle cell layer of epithelial dysplasia, 80KD leaved product was expressed more strongly.

4. The expression of RAD21 cleaved products by ALLN

The result suggested that RAD21 cleaved products related to differentiation of oral squamous epithelium. Then, we investigated the relation between the cleavage of RAD21 and calpain, known as a cystein protease related to differentiation of cell. SAS cells were treated with 10uM ALLN, a calpain inhibitor, and/or 10uM 5-azacytidine for 24 hours. Control cells were incubated with condition medium. Each cell was fractionated to cytoplasm and nucleus, and analysed RAD21 cleaved products. As a result, we could detect 80KD product increased in cytoplasm with 10uM 5-azacytidine. But, the product decreased with 5-azacytidine and ALLN (Fig.7).

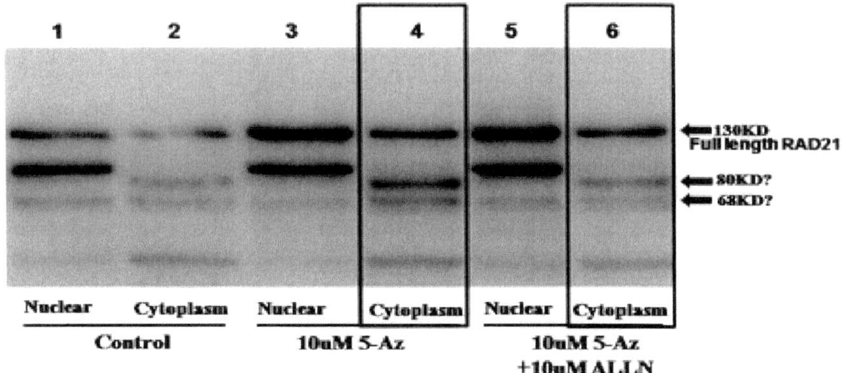

Fig.7 Expression RAD21 cleaved products in SAS cells. Treatment with 5-azacytidine and ALLN. 80KD product increased in cytoplasm with 10uM 5-azacytidine(lane No.4). But, the product decreased with 5-azacytidine and ALLN(Lane No.6).

Discussion

We studied on the relation between the invasion pattern of squamous cell carcinoma and the expression level of the RAD21 gene using the Laser microdissection method and real-time PCR in surgical material. It became clear that the expression level of the RAD21 gene decreased as the invasion pattern of squamous cell carcinoma changed to INFγ from INFα. It is known that the expression level of the RAD21 gene is down-regulated under hypoxia in many human tumor cells (Sook Kim et al. 2001). As well, hypoxia in malignant tumor tissue is more concerned in development of the aggressive character and metastasis of malignant tumor cells (Acs et al. 2002, Beavon 1999, 2000, Brizel et al. 1996, Cairns et al. 2001, Schwickert et al. 1995). It has been shown that the lowering of oxygen partial pressure in the tumor correlates with prognosis in squamous cell carcinoma of the head and neck (Brizel et al. 1997). In regard to the biological significance of the invasion pattern, it is known that in oral squamous cell carcinoma, similar to cancer in other organs, an invasive growth pattern such as INFγ indicates high lymph node metastasis frequency and poor prognosis, as compared with the cancer which shows an expansive growth pattern such as INFα (Baba et al. 1997, Kumagai et al. 1994, Nishida et al. 1995). Based on the above reports and the results of our study, the RAD21 gene might be related to the development of an invasive pattern. Simultaneously, these results suggest the possibility that the invasion pattern of squamous cell carcinoma reflects the hypoxia in the tumor tissue.

In regional lymph node metastasis, the expression level of the RAD21 gene decreased similary to INFγ invasion pattern. This result supports the possibility that cancer cells with high invasion potential cause lymph node metastasis in relation to hypoxic condition.

The RAD21 gene has been identified, by means of the differential display method, as the gene which is down-regulated in the hypoxia (Sook Kim et al. 2001). This suggests the possibility that the decrease of the expression level of the RAD21 gene affects various intracellular changes in the hypoxia. In solid tumors, hypoxic condition concerns a selection of the apoptosis resistant cells, and it has been indicated to be important for progression of tumor cells (Graeber et al. 1996). As well, hypoxia directly induces the cell's apoptotic resistance (Baek et al. 2000, Dong et al. 2003, Volm and Koomagi 2000). In addition, it has been shown that the angiogenesis of cancer and the apoptosis of cancer cells correlate inversely in squamous cell carcinoma of the head and neck. These are considered to be the result of apoptotic resistance induced by hypoxia in cancer cells (Riedel et al. 2001).

RAD21 is cleaved by caspase 3, and it has become clear that the degradation product in the C-terminal site induces apoptosis, so the RAD21 gene plays an important role in apoptosis (Chen et al. 2002, Pati et al. 2002). These reports suggest that the apoptotic resistance of the tumor cell and the down-regulation of the RAD21 gene are closely related. But, in this study, we detected new cleaved product of RAD21. That product was different from cleavage product by caspase. It is known that there were many cleavage product of RAD21(Chen et al. 2002). Though, it is unclear what protease cleaves RAD21. Our data indicate at least calpain may cleaves RAD21 and that cleaved product correlate with cell differentiation. Finally, our findings suggest that the RAD21 gene plays an important role in the control of cell differentiation, apoptosis and growth of cancer cells.

Acknowledgements

This work was supported in part by a Showa University Grant-in-Aid for Innovative Collaborative Research Projects, the High-Technology Research Center Project, Characteristic Education from the Ministry of Education, Culture, Sports, Science and Technology of Japan and Smoking Research Foundation.

References

Acs G, Zhang PJ, Rebbeck TR, Acs P,Verma A (2002) Immunohistochemical expression of erythropoietin and erythropoietin receptor in breast carcinoma. Cancer 95: 969-981

Atalay A, Crook T, Ozturk M,Yulug IG (2002) Identification of genes induced by BRCA1 in breast cancer cells. Biochem Biophys Res Commun 299: 839-846

Baba H, Ohshiro T, Yamamoto M, Endo K, Adachi E, Kakeji Y, Kohnoe S, Maehara Y,Sugimachi K (1997) Clinicopathological characteristics of stage 1 gastric cancer: comparison of macroscopic and microscopic findings. Hepatogastroenterology 44: 554-558

Baek JH, Jang JE, Kang CM, Chung HY, Kim ND,Kim KW (2000) Hypoxia-induced VEGF enhances tumor survivability via suppression of serum deprivation-induced apoptosis. Oncogene 19: 4621-4631

Beavon IR (1999) Regulation of E-cadherin: does hypoxia initiate the metastatic cascade? Mol Pathol 52: 179-188

Beavon IR (2000) The E-cadherin-catenin complex in tumour metastasis: structure, function and regulation. Eur J Cancer 36: 1607-1620

Birkenbihl RP,Subramani S (1992) Cloning and characterization of rad21 an essential gene of Schizosaccharomyces pombe involved in DNA double-strand-break repair. Nucleic Acids Res 20: 6605-6611

Birkenbihl RP,Subramani S (1995) The rad21 gene product of Schizosaccharomyces pombe is a nuclear, cell cycle-regulated phosphoprotein. J Biol Chem 270: 7703-7711

Brizel DM, Scully SP, Harrelson JM, Layfield LJ, Bean JM, Prosnitz LR,Dewhirst MW (1996) Tumor oxygenation predicts for the likelihood of distant metastases in human soft tissue sarcoma. Cancer Res 56: 941-943

Brizel DM, Sibley GS, Prosnitz LR, Scher RL,Dewhirst MW (1997) Tumor hypoxia adversely affects the prognosis of carcinoma of the head and neck. Int J Radiat Oncol Biol Phys 38: 285-289

Cairns RA, Kalliomaki T,Hill RP (2001) Acute (cyclic) hypoxia enhances spontaneous metastasis of KHT murine tumors. Cancer Res 61: 8903-8908

Chen F, Kamradt M, Mulcahy M, Byun Y, Xu H, McKay MJ,Cryns VL (2002) Caspase proteolysis of the cohesin component RAD21 promotes apoptosis. J Biol Chem 277: 16775-16781

Dong Z, Wang JZ, Yu F,Venkatachalam MA (2003) Apoptosis-resistance of hypoxic cells: multiple factors involved and a role for IAP-2. Am J Pathol 163: 663-671

Graeber TG, Osmanian C, Jacks T, Housman DE, Koch CJ, Lowe SW,Giaccia AJ (1996) Hypoxia-mediated selection of cells with diminished apoptotic potential in solid tumours. Nature 379: 88-91

Hauf S, Waizenegger IC,Peters JM (2001) Cohesin cleavage by separase required for anaphase and cytokinesis in human cells. Science 293: 1320-1323

Hirano T (2000) Chromosome cohesion, condensation, and separation. Annu Rev Biochem 69: 115-144

Hoque MT, Ishikawa F (2001) Human chromatid cohesin component hRad21 is phosphorylated in M phase and associated with metaphase centromeres. J Biol Chem 276: 5059-5067

Kumagai S, Kojima S, Imai K, Nakagawa K, Yamamoto E, Kawahara E,Nakanishi I (1994) Immunohistologic distribution of basement membrane in oral squamous cell carcinoma. Head Neck 16: 51-57

Michaelis C, Ciosk R,Nasmyth K (1997) Cohesins: chromosomal proteins that prevent premature separation of sister chromatids. Cell 91: 35-45

Nasmyth K, Peters JM,Uhlmann F (2000) Splitting the chromosome: cutting the ties that bind sister chromatids. Science 288: 1379-1385

Nishida T, Tanaka S, Haruma K, Yoshihara M, Sumii K,Kajiyama G (1995) Histologic grade and cellular proliferation at the deepest invasive portion correlate with the high malignancy of submucosal invasive gastric carcinoma. Oncology 52: 340-346

Pati D, Zhang N,Plon SE (2002) Linking sister chromatid cohesion and apoptosis: role of Rad21. Mol Cell Biol 22: 8267-8277

Porkka KP, Tammela TL, Vessella RL,Visakorpi T (2004) RAD21 and KIAA0196 at 8q24 are amplified and overexpressed in prostate cancer. Genes Chromosomes Cancer 39: 1-10

Riedel F, Gotte K, Bergler W,Hormann K (2001) Inverse correlation of apoptotic and angiogenic markers in squamous cell carcinoma of the head and neck. Oncol Rep 8: 471-476

Schwickert G, Walenta S, Sundfor K, Rofstad EK,Mueller-Klieser W (1995) Correlation of high lactate levels in human cervical cancer with incidence of metastasis. Cancer Res 55: 4757-4759

Sonoda E, Matsusaka T, Morrison C, Vagnarelli P, Hoshi O, Ushiki T, Nojima K, Fukagawa T, Waizenegger IC, Peters JM, Earnshaw WC,Takeda S (2001) Scc1/Rad21/Mcd1 is required for sister chromatid cohesion and kinetochore function in vertebrate cells. Dev Cell 1: 759-770

Sook Kim M, Hyen Baek J, Bae MK,Kim KW (2001) Human rad21 gene, hHR21(SP), is downregulated by hypoxia in human tumor cells. Biochem Biophys Res Commun 281: 1106-1112

Takahashi K, Kanazawa H, Akiyama Y, Tazaki S, Takahara M, Muto T, Tanzawa H,Sato K (1989) Establishment and characterization of a cell line (SAS) from poorly differentiated human squamous cell carcinoma of the tongue. J Jpn Stomatol Soc 38: 20-28

Uhlmann F, Lottspeich F,Nasmyth K (1999) Sister-chromatid separation at anaphase onset is promoted by cleavage of the cohesin subunit Scc1. Nature 400: 37-42

Volm M,Koomagi R (2000) Hypoxia-inducible factor (HIF-1) and its relationship to apoptosis and proliferation in lung cancer. Anticancer Res 20: 1527-1533

Telomerase-Specific Replication-Selective Virotherapy for Oral Squamous Cell Carcinoma Cell Lines

Yuji Kurihara[1], Yuichi Watanabe[3,4], Toru Kojima[2,3], Tatsuo Shirota[1], Masashi Hatori[1], Yasuo Urata[4], Toshiyoshi Fujiwara[2,3], and Satoru Shintani[1]

[1]Department of Oral and Maxillofacial Surgery, School of Dentistry, Showa University, Tokyo, Japan
[2]Division of Surgical Oncology, Department of Surgery, Okayama University, Graduate School of Medicine and Dentistry, Okayama, Japan
[3]Center for Gene and Cell Therapy, Okayama University Hospital, Okayama, Japan
[4]Oncolys BioPharma, Inc., Tokyo, Japan

Summary. Replication-selective tumor-specific viruses present a novel approach for treating neoplastic disease. These vectors are designed to induce virus-mediated lysis of tumor cells after selective viral propagation within the tumor. Telomerase activation is considered to be a critical step in carcinogenesis, and its activity is closely correlated with human telomerase reverse transcriptase (hTERT) expression. We investigated the antitumor effect of the hTERT-specific replication-competent adenovirus on oral squamous cell carcinoma (OSCC) cells. An adenovirus 5 vector [tumor- or telomerase-specific replication-competent adenovirus (OBP-301)] was constructed by Center for Gene and Cell Therapy, Okayama University Hospital, in which the hTERT promoter element drives expression of *E1A* and *E1B* genes linked with an internal ribosome entry site, and we examined the selective replication and antitumor effect in human OSCC cells *in vitro* and *in vivo* orthotopic graft model. OBP-301 replicated efficiently and induced more over 50% cell killing in a panel of human OSCC cell lines, but not in normal human fibroblasts, which were lacking telomerase activity. In orthotopic graft model, intratumoral injection of OBP-301 resulted in a significant inhibition of tumor growth and prolongation of survival. Furthermore, mice's weight recovered as same as before bearing OSCC tumors. Our data clearly indicate that OBP-301 displays an acceptable toxicity profile as well as a therapeutic oncolytic activity for OSCC cells *in vitro* and *in vivo*. These results suggested that treatment of OBP-301 is possible to improvement of QOL of oral cancer patients.

Key words. oral cancer; oncolysis; adenovirus; telomerase

Introduction

Cancer remains a leading cause of death worldwide despite improvements in diagnostic techniques and clinical management (Parkin et al. 2005, Jemal et al. 2004). Squamous cell carcinoma of the head and neck (HNSCC) distresses an estimated 500,000 patients annually worldwide. This aggressive epithelial malignancy is associated with a high mortality and severe morbidity among the long-term survivors (Vokes et al. 1993). Current treatment strategies for advanced cancer include surgical resection, radiation, and cytotoxic chemotherapy. Although a combination of these modalities can improve survival, most patients die of disease progression, often resulting from intrinsic or acquired resistance to treatment (Vokes et al. 2005, Milas et al. 2003). Lack of specificity for tumor cells is the primary limitation of radiotherapy and chemotherapy. To improve the therapeutic index, there is a need for anticancer agents that selectively target tumor cells only and spare normal cells.

Replication-selective tumor-specific viruses present a novel approach for cancer treatment. These vectors are designed to induce virus-mediated lysis of tumor cells after selective viral propagation within the tumor tissue only (Kirn et al. 2001, Hawkins et al. 2002). Targeting cancer cells effectively also requires tissue- or cell-specific promoters that can express in diverse tumor types and are silent in normal cells.

Telomerase is a ribonucleoprotein complex responsible for the complete replication of chromosomal ends (Blackburn 1991). Telomerase activity is present in more than 85% of human cancers (Kim et al. 1994), but only in few normal somatic cells (Shay et al. 1996). Telomerase activation is considered a critical step in carcinogenesis and its activity is closely correlated with human telomerase reverse transcriptase (*hTERT*) expression (Nakayama et al. 1998). Since only tumor cells that express telomerase activity would activate this promoter, the *hTERT* proximal promoter allows for preferential expression of viral genes in tumor cells, leading to selective viral replication.

Center for Gene and Cell Therapy, Okayama University Hospital constructed an attenuated adenovirus 5 vector (OBP-301; Telomelysin), in which the *hTERT* promoter element drives the expression of *E1A* and *E1B* genes linked to an internal ribosome entry site (IRES) (Kawashima et al. 2004). Telomelysin replicated efficiently and induced marked cell killing of a panel of human cancer cell lines, whereas replication as well as cytotoxicity was highly attenuated in normal human cells lacking telomerase activity.

An orthotopic graft nude mouse model of oral tongue squamous cell carcinoma resembles human HNSCC in a number of biological properties. In this model, local tumor growth, and regional and distant metastases demonstrated a histopathological similarity to Oral squamous cell carcinoma (OSCC) primary tumors from patients (Myers et al. 2002). In the present study, we examined the selective replication and antitumor effects in human OSCC cells such as in vitro oncolysis and a therapeutic benefit *in vivo*.

Materials and Methods

Construction of OBP-301

A 897-bp fragment of the E1A gene was amplified by reverse transcription-PCR from total cellular RNA of 293 cells using the primers E1A-S (5'-ACA-CCG-GGA-CTG-AAA-ATG-AG-3') and E1A-AS (5'-CAC-AGG-TTT- ACA-CCT-TAT-GGC-3'). A 1822-bp fragment of the E1B gene was amplified by PCR from genomic DNA of 293 cells using the primers E1B-S (5'-CTG-ACC-TCA- TGG-AGG-CTT-GG-3') and E1B-AS (5'-GCC-CAC-ACA-TTT-CAG-TAC-CTC-3'). The amplified products were subcloned into the pTA plasmid according to the instructions provided by the manufacturer (Invitrogen, Carlsbad, CA). After confirmation by DNA nucleotide sequencing, the *E1A* gene (911 bp) and the *E1B* gene (1836 bp) were digested with EcoRI and then cloned into the pIRES vector (Clontech, Palo Alto, CA) at the *Mlu*I and *Sal*I restriction sites, respectively (pE1A-IRES-E1B). A 455-bp fragment of the hTERT promoter was digested with *Mlu*I and *Bgl*II restriction enzymes from pGL3–378, which contains a 378-bp region upstream of the transcription start site, and then ligated into the *Xho*I site of the pE1A-IRES-E1B (phTERT-E1A-IRES-E1B). The 3828-bp fragment was digested from phTERT-E1A-IRES-E1B with *Nhe*I and *Not*I restriction enzymes and then cloned into pShuttle after deletion of the cytomegalovirus promoter. The resultant shuttle vector was used to generate replication-competent adenovirus under control of the *hTERT* promoter using the Adeno-X Expression System (Clontech) according to the protocol provided by the manufacturer. Recombinant adenovirus was isolated from a single plaque and expanded in 293A cells. PCR was performed using several primers specific for adenovirus *E1* sequences, and the PCR products were digested with restriction enzymes to confirm the structure. The E1A-deleted adenovirus vector d1312 was used as the control vector. Recombinant viruses were purified by ultracentrifugation in cesium chloride step gradients, and their titers were determined by plaque assay in the 293 cells.

Cell lines and cell culture

Oral squamous carcinoma cell lines (SAS-L, HSC-2, HSC-3, HSC-4) were maintained as monolayers in Dulbecco's modified Eagle's medium (DMEM) supplemented with 10% heat-inactivated fetal bovine serum (FBS), 100 units/ml penicillin and 100 mg/ml streptomycin (complete medium). H1299, the human non-small-cell lung cancer cell line, and LNcap, the human prostate cancer cell line were routinely propagated in monolayer culture in RPMI-1640 supplemented with 10% FBS.

Cell viability assay

XTT (sodium 3'-[1-(phenylaminocarbonyl)-3,4- tetrazolium] -bis (4-methoxy-6 -nitro) benzene sulfonic acid hydrate) assays were performed by seeding human OSCC cells at 1,000 cells/well in 96-well plates 18-20 hr before viral infection. The cells were then infected with OBP-301 at a multiplicity of infection (MOI) of 1, 10, 50 and 100 plaque-forming units (PFU)/cell. Cell viability was determined at the indicated times by using a Cell Proliferation Kit II (Roche Molecular Biochemicals, Indianapolis, IN) according to the protocol provided by the manufacturer.

Quantitative real-time PCR assay

Cells were seeded on 6 well plates at 5 x 10^4 cells, 18-20 hours before infection. Cells were infected with OBP-301 at an MOI of 5 PFU/cell for 2 hours. Following the removal of virus, the cells were incubated at 37 °C, trypsinized, and harvested for intracellular replication analysis at 2, 24, 48, and 72 hours. DNA was extracted with QIAamp DNA Mini Kit (Qiagen Inc., Valencia, CA), and quantitative real-time PCR assay for the *E1A* gene was performed using a LightCycler instrument (Roche Molecular Biochemicals). The sequences of specific primers used for *E1A* were as follows: sense, 5'-

CCT GTG TCT AGA GAA TGC AA-3' and antisense, 5'-ACA GCT CAA GTC CAA AGG TT-3'. PCR amplification began with a 600-sec denaturation step at 95°C and then 40 cycles of denaturation at 95°C for 10 sec, annealing at 58°C for 15 sec, and extension at 72°C for 8 sec. Data were analyzed using LightCycler Software (Roche Molecular Biochemicals). The ratios normalized by dividing the value of untreated cells were presented for each sample.

Total RNA from cultured cells was obtained using the RNeasy Mini Kit (Qiagen, Chatsworth, CA). Approximately 0.1 μg of total RNA was used for reverse transcription. Reverse transcription was performed at 22 °C for 10 minutes and then at 42 °C for 20 minutes. The *hTERT* mRNA copy number was determined by real-time quantitative reverse transcription-polymerase chain reaction (RT-PCR) using a LightCycler instrument and a LightCycler DNA TeloTAGGG Kit (Roche Molecular Biochemicals, Indianapolis, IN). Polymerase chain reaction amplification began with a 60-second denaturation step at 95 °C, followed by 40 cycles of denaturation at 95 °C for 15 seconds, annealing at 58 °C for 10 seconds, and extension at 72 °C for 9 seconds.

Flow cytometry

The cells (2×10^5 cells) were labeled with mouse monoclonal anti-CAR (RmcB; Upstate Biotechnology, NY) for 30 min at 4 °C, incubated with FITC-conjugated rabbit anti-mouse IgG second antibody (Zymed Laboratories, San Francisco), and analyzed by the FACSCalibur (Becton Dickinson, Mountain View, CA) using CELL Quest software. The window was set to exclude dead cells and debris.

Orthotopic graft model of tongue cancer

Female BALB/c nude mice were used for the animal model of tongue cancer (Clea Japan, Inc., Tokyo, Japan). They were 6-week-old and weighted between 15 and 20 g. A cell line of human oral squamous cell carcinoma, SAS-L cells were harvested and suspended at a concentration of 5×10^6 cells/ml in Dulbecco's modified Eagle's medium (DMEM). A total amount of 0.02 ml cell suspension (1×10^5 cells) was injected into the right lateral border of tongue of the nude mice with a 27-gauge needle. About 5-7 days later, the tumors grew to 2–3 mm in diameter. At that time, a 0.02 ml solution containing OBP-301 at a dose of 1×10^8 PFU/body or PBS was injected into the tumor. The perpendicular diameter of each tumor was measured every 3 days, and tumor volume was calculated using the following formula: tumor volume (mm^3) $=a \times b^2 \times 0.5$, where a is the longest diameter, b is the shortest diameter, and 0.5 is a constant to calculate the volume of an ellipsoid. Body weight of each mouse was measured on every day. The experimental protocol was approved by the Ethics Review Committee for Animal Experimentation of Okayama University.

Results

Construction of adenovirus containing hTERT-controlled *E1A* and *E1B* genes

To construct an adenovirus that replicates selectively in telomerase-positive cells, the *E1A* and *E1B* genes linked with the internal ribosome entry site (IRES) were placed under the control of the *hTERT* promoter. OBP-301 was constructed by inserting this expression cassette at the deleted E1 region of the replication-deficient adenovirus type 5 virus (Fig.1).

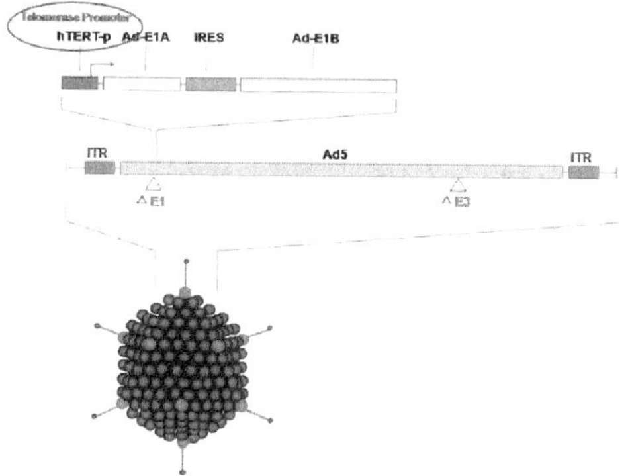

Fig.1. A schematic DNA structure of tumor- or telomerase-specific replication-competent adenovirus. Tumor- or telomerase-specific replication-competent adenovirus contains the human telomerase reverse transcriptase (*hTERT*) promoter sequence inserted into the E3-deleted adenovirus genome to drive transcription of the *E1A* and *E1B* bicistronic cassette linked by the internal ribosome entry site structure.

Human telomerase reverse transcriptase levels in oral cancer and normal cells

Telomerase is a novel marker for malignant diseases (Shay and Wright 2002). To confirm the specificity of telomerase activity in human cancer cells, expression of *hTERT* mRNA, which plays a key role in telomerase activation (Nakayama et al. 1998) was measured in a panel of human cancer cells and normal cell lines using a real-time RT-PCR method. All tumor cell lines, including human non–small-cell lung cancer cell line H1299 and human oral cancer cell lines SAS-L, SAS-L1 GFP, and HSC-3, expressed detectable levels of *hTERT* mRNA, although the levels of expression varied widely (Fig.2). In contrast, human fibroblast cell such as WI-38 was negative for *hTERT* expression. These results suggest that the *hTERT* promoter element can be used to target for human cancer cells.

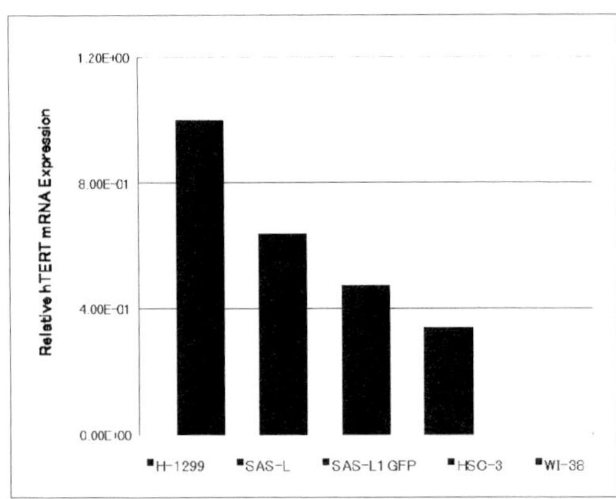

Fig.2. Relative *hTERT* mRNA expression in human tumor and normal cells were determined by real-time RT-PCR analysis. The *hTERT* mRNA expression of H1299 human lung cancer cells was considered to be 1.0, and the relative expression level of each cell line was calculated against that of H1299 cells.

The expression of coxackie-adenovirus receptor (CAR) in human oral cancer cells

As shown in Fig.3, to determine on the cell surface expression of CAR in human oral cancer cells, we used flow cytometric analysis. The expression of

CAR in human oral cancer cells, SAS-L and HSC-3, were almost same level compared with these of H1299, the human non-small-cell lung cancer cell line. These results suggest that there is no difference in the expression of CAR between OSCC and other cancer cells.

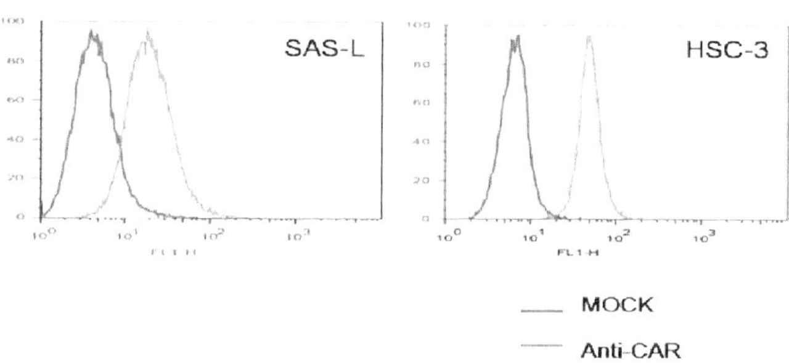

Fig.3. Expression of CAR on OSCC cell lines as determined by flow cytometric analysis. The cells (2×10^5 cells) were labeled with mouse monoclonal anti-CAR for 30 min at 4 °C, incubated with FITC-conjugated rabbit anti-mouse IgG second antibody.

In vtro cytotoxic effects of OBP-301 in human oral cancer cells

To determine whether OBP-301 infection induces selective cell lysis, four oral cancer cell lines (SAS-L, HSC-2, HSC-3, and HSC-4) were infected with OBP-301 at various MOIs. All oral cancer cell lines were more than 50% killed by OBP-301 in a dose-dependent fashion, but not in normal human fibroblasts, which were lacking telomerase activity (Data not shown). (Fig. 4).

Replication selectivity of OBP-301 in human oral cancer cells

To determined replication efficiency of OBP-301 in human oral cancer (SAS-L) and normal human lung fibroblast (MRC-5) by real-time PCR assay. Twenty four-well plates were seeded with 5×10^4 cells/well 24 h before infection. Cells were infected with OBP-301 at a MOI of 10 PFU/cell. Virus inocula were removed after 2 h. The cells were washed twice with medium, and 1 ml of the medium was added to each well. The cells were incubated at 37 °C for varying periods of time. The transduction efficiency of adenovirus is less efficient in normal cells compared with tumor cell (Fig. 5).

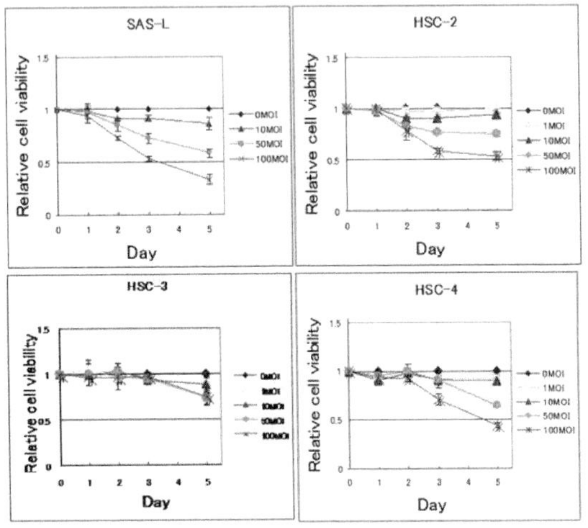

Fig.4. Oncolytic efficacy induced by OBP-301 infection in vitro. Cell killing efficacy was evaluated by XTT (sodium 3'-[1-(phenylaminocarbonyl)-3,4- tetrazolium] -bis (4-methoxy-6 -nitro) benzene sulfonic acid hydrate) assays. Statistical analysis was performed using Student's t test for differences among groups. Statistical significance (*) was defined as $P < 0.05$.

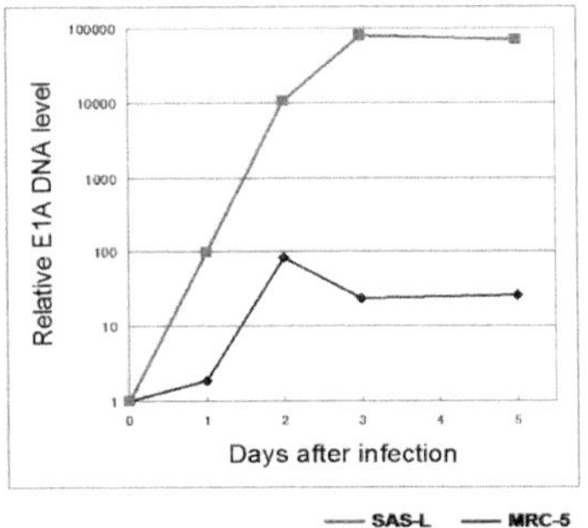

Fig.5. To determined replication efficiency of OBP-301 in human oral cancer (SAS-L) and normal human lung fibroblast (MRC-5) by real-time PCR assay. Twenty four-well plates were seeded with 5 x 10^4 cells/well 24 h before infection. Cells were infected with OBP-301 at a MOI of 10 PFU/cell. Virus inocula were removed after 2 h The cells were washed twice with medium, and 1 ml of the medium was added to each well. The cells were incubated at 37°C for varying periods of time. The transduction efficiency of adenovirus is less efficient in normal cells compared with tumor cell.

Treatment of Human Tumor Xenografts with Intratumoral Injection of OBP-301

We next assessed the therapeutic efficacy of OBP-301 against human oral cancer cells *in vivo*. SAS-L cells were implanted as xenografts into the tongue of BALB/c *nu/nu* mice. Mice bearing palpable SAS-L tumors with a diameter of 3–5 mm received three daily courses of intratumoral injection of 10^8 PFU of OBP-301 or PBS (mock treatment). Compared with PBS, tumor growth was suppressed by OBP-301 injection. And the weight of mice treated with OBP-301 was significantly recovered compared with PBS (Fig. 6-A). The macroscopic appearance showed that tumors treated with OBP-301 were consistently smaller than those of other cohorts of mice 28 days after treatment (Fig. 6-B). And mice treated with OBP-301 had a significant increased in survival (Fig. 6-C). These *in vivo* data closely correlate with the *in vitro* results.

Discussion

The present study illustrates the potential application of replication-selective adenovirus as an anticancer agent. Based on knowledge of the adenoviral replication cycle, there are two types of approaches to restrict its replication to tumor cells. ONYX-015 is an adenovirus with a deleted *E1B* 55-kDa gene, which replicates and causes cell killing in p53-deficient tumor cells, but not in normal cells (Bischoff et al. 1996). However, many studies have demonstrated that replication of ONYX-015 is not strictly linked to the deficiency of p53 and that ONYX-015 also replicates in normal human primary cells (Rothmann et al. 1998, Geoerger et al. 2002). An alternate strategy to obtain tumor-specific adenoviral replication has been developed by using heterologous promoters that regulate transcription of the E1A gene. In this context, the promoters from the prostate-specific antigen (Rodriguez et al. 1997), MUC1 (Kurihara et al. 2000), osteocalcin (Matsubara et al. 2001), L-plastin (Peng et al. 2001), midkine (Adachi et al. 2001), and E2F-1 (Tsukuda et al. 2002) genes have been used to drive *E1A* expression. These vectors replicate preferentially in tumor cells that express each targeted tumor marker; their therapeutic window, however, is relatively narrow because only part of the tumor is positive for each tumor marker.

Telomerase is expressed in a majority of human cancers, and its activation plays a critical role in tumorigenesis by sustaining cellular immortality (Kim et al. 1994), suggesting that the *hTERT* promoter is preferentially activated in

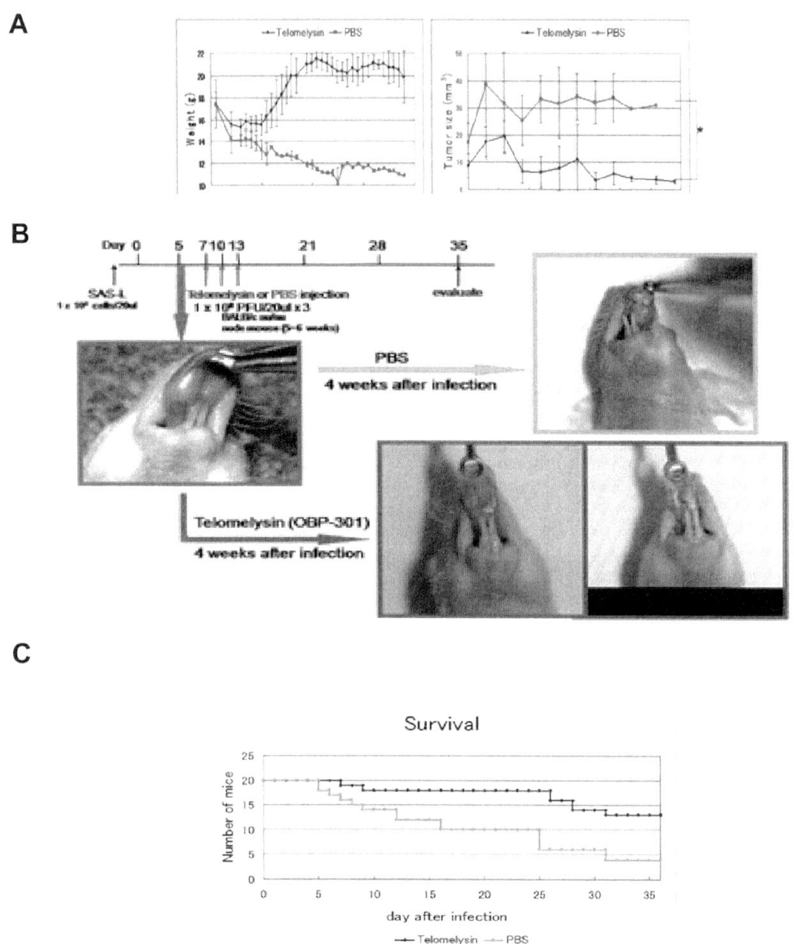

Fig.6. Effect of tumor- or telomerase-specific replication-competent adenovirus on human cancer cells grown in nu/nu mice. A. antitumor effects of intratumorally injected tumor- or telomerase-specific replication-competent adenovirus against established tongue SAS-L xenograft tumors in nu/nu mice. PBS was used as a control. Eight mice were used for each group. Tumor growth was expressed as mean tumor volume ± SD. Statistical significance (*) was defined as $P < 0.05$ (Student's t test). B. macroscopic appearance of SAS-L tongue tumors in nu/nu mice at 0 and 28 days after treatment. C. Mice treated with OBP-301 had a significant increased in survival. Refer to color plates.

most human cancer cells. Thus, the broadly applicable *hTERT* promoter might be a suitable regulator of adenoviral replication. In fact, it has been reported recently that transcriptional control of *E1A* expression via the *hTERT* promoter could restrict adenoviral replication to telomerase-positive tumor cells and efficiently lyse tumor cells (Wirth et al. 2003). The *E1A* gene has been shown to have tumor-suppressive activity including transcriptional repression of the HER-2/neu proto-oncogene and induction of apoptosis, suggesting that *E1A* gene transfer alone might be sufficient to kill host cells (Deng et al. 2002). The adenovirus *E1B* gene is expressed early in viral infection, and its gene product inhibits *E1A*-induced p53-dependent apoptosis, which in turn promotes the cytoplasmic accumulation of late viral mRNA, leading to shutoff of host cell protein synthesis. In most vectors that replicate under the transcriptional control of the *E1A* gene, the *E1B* gene is driven by the endogenous adenovirus *E1B* promoter. Recent studies have demonstrated that transcriptional control of both *E1A* and *E1B* genes by the α-fetoprotein promoter with the use of IRES significantly improved the specificity and therapeutic index in hepatocellular carcinoma cells (Li et al. 2001). These observations indicate that the OBP-301, in which the *hTERT* promoter regulates both the *E1A* and *E1B* genes, may control viral replication more stringently, thereby providing profound therapeutic effects in tumor cells as well as attenuated toxicity in normal tissues.

Kawashima *et al.* reported that an attenuated adenovirus OBP-301 could replicate in and causes selective lysis of human cancer cells (Kawashima et al. 2004). These reports indicated that tumor implanted as xenografts into the hind flank of nu/nu mice was significantly suppressed by the infection of OBP-301. In this study, we examined the effects of OBP-301 on oral squamous cell carcinoma cells, in which used orthotopic graft model of tongue cancer. This model would be expected that not only suppress the tumor growth in their tongue but also recover the efficacy of eating and swallowing. Because we thought that recovered of these actions of mice leads to the improvement of Quality of Life (QOL) of oral cancer patients.

An orthotopic nude mouse model to investigate the cellular and molecular mechanisms of metastasis in human neoplasia was first described by Fidler *et al.* (Fidler 1986, Fidler et al. 1990) and Killion *et al.* (Killion et al. 1998). Other groups using human solid tumors from a variety of organ sites have subsequently documented its significance (Tan et al. 1977, Hoffman 1994). The orthotopic implantation of tumor cells restores the correct tumor-host interactions, which do not occur when tumors are implanted in ectopic s.c. sites (Fidler 1986). In addition, an accirate experimental animal model is necessary to evaluate the efficacy of novel therapies (Killion et al. 1998,

Hoffman 1994). The first report of orthotopic model of human oral cancer in nude mice was published by Fitch *et al.* (Fitch et al. Wolf G. T., Carey T. E. 1988), who aspirated cells from fresh human tumors growing s.c. in nude mice and injected them into the tongues of nude mice. These studies showed an equal tumorigenicity in the oral cavity and the subcutis. Although these reports of oral cancer orthotopic graft models were shown, there are no reports that the suggestion of the improvement of QOL by the treatment of therapeutic reagents for cancer.

Our data showed that human OSCC cell lines SAS-L, SAS-L1 GFP, and HSC-3, expressed detectable levels of *hTERT* mRNA, although the levels of expression varied widely. In contrast, human fibroblast cell such as WI-38 was negative for *hTERT* expression. We also showed cytotoxic activity in a variety of human OSCC cell lines. Furthermore, we demonstrated that tumor growth of orthotopic graft model was significantly suppressed by the treatment of OBP-301. These results suggest that OBP-301 exhibited desirable features for use as an oncolytic therapeutic agent. OBP-301 displayed potent lytic activity in OSCC cells both in vitro and in vivo. Especially, in oral cancer orthotopic graft model, intratumoral injection of OBP-301 resulted in a significant inhibition of tumor growth and survival. Furthermore, mice's weight recovered as same as before implanted oral cancer cells into their tongues. We suggested that the reduction of the tongue tumor by the treatment of OBP-301 is enabled to recover the efficacy of eating and swallowing well again. These results suggested that treatment of OBP-301 is possible to improvement of QOL of oral cancer patients.

In conclusion, our data clearly indicate that OBP-301 displays an acceptable toxicity profile as well as a therapeutic oncolytic activity for oral tumors. Additional clinical studies will be required to address the issues of safety and efficacy of OBP-301-mediated virotherapy for human oral cancers.

References

Adachi Y, Reynolds PN, Yamamoto M, Wang M, Takayama K, Matsubara S, Muramatsu T,Curiel DT (2001) A midkine promoter-based conditionally replicative adenovirus for treatment of pediatric solid tumors and bone marrow tumor purging. Cancer Res 61: 7882-7888

Bischoff JR, Kirn DH, Williams A, Heise C, Horn S, Muna M, Ng L, Nye JA, Sampson-Johannes A, Fattaey A,McCormick F (1996) An adenovirus mutant that replicates selectively in p53-deficient human tumor cells. Science 274: 373-376

Blackburn EH (1991) Structure and function of telomeres. Nature 350: 569-573

Deng J, Kloosterbooer F, Xia W,Hung MC (2002) The NH(2)-terminal and conserved region 2 domains of adenovirus E1A mediate two distinct mechanisms of tumor suppression. Cancer Res 62: 346-350

Fidler IJ (1986) Rationale and methods for the use of nude mice to study the biology and therapy of human cancer metastasis. Cancer Metastasis Rev 5: 29-49

Fidler IJ, Naito S,Pathak S (1990) Orthotopic implantation is essential for the selection, growth and metastasis of human renal cell cancer in nude mice [corrected]. Cancer Metastasis Rev 9: 149-165

Fitch KA, Somers KD, Schechter GL (1988) The development of a head and neck tumor model in the nude mouse. In: Wolf G. T., Carey T. E. (ed) Head and Neck Oncology Research. Kugler Publications, Amsterdam, pp187-190

Geoerger B, Grill J, Opolon P, Morizet J, Aubert G, Terrier-Lacombe MJ, Bressac De-Paillerets B, Barrois M, Feunteun J, Kirn DH,Vassal G (2002) Oncolytic activity of the E1B-55 kDa-deleted adenovirus ONYX-015 is independent of cellular p53 status in human malignant glioma xenografts. Cancer Res 62: 764-772

Hawkins LK, Lemoine NR,Kirn D (2002) Oncolytic biotherapy: a novel therapeutic plafform. Lancet Oncol 3: 17-26

Hoffman RM (1994) Orthotopic is orthodox: why are orthotopic-transplant metastatic models different from all other models? J Cell Biochem 56: 1-3

Jemal A, Clegg LX, Ward E, Ries LA, Wu X, Jamison PM, Wingo PA, Howe HL, Anderson RN,Edwards BK (2004) Annual report to the nation on the status of cancer, 1975-2001, with a special feature regarding survival. Cancer 101: 3-27

Kawashima T, Kagawa S, Kobayashi N, Shirakiya Y, Umeoka T, Teraishi F, Taki M, Kyo S, Tanaka N,Fujiwara T (2004) Telomerase-specific replication-selective virotherapy for human cancer. Clin Cancer Res 10: 285-292

Killion JJ, Radinsky R,Fidler IJ (1998) Orthotopic models are necessary to predict therapy of transplantable tumors in mice. Cancer Metastasis Rev 17: 279-284

Kim NW, Piatyszek MA, Prowse KR, Harley CB, West MD, Ho PL, Coviello GM, Wright WE, Weinrich SL,Shay JW (1994) Specific association of human telomerase activity with immortal cells and cancer. Science 266: 2011-2015

Kirn D, Martuza RL,Zwiebel J (2001) Replication-selective virotherapy for cancer: Biological principles, risk management and future directions. Nat Med 7: 781-787

Kurihara T, Brough DE, Kovesdi I,Kufe DW (2000) Selectivity of a replication-competent adenovirus for human breast carcinoma cells expressing the MUC1 antigen. J Clin Invest 106: 763-771

Li Y, Yu DC, Chen Y, Amin P, Zhang H, Nguyen N,Henderson DR (2001) A hepatocellular carcinoma-specific adenovirus variant, CV890, eliminates distant human liver tumors in combination with doxorubicin. Cancer Res 61: 6428-6436

Matsubara S, Wada Y, Gardner TA, Egawa M, Park MS, Hsieh CL, Zhau HE, Kao C, Kamidono S, Gillenwater JY,Chung LW (2001) A conditional replication-competent adenoviral vector, Ad-OC-E1a, to cotarget prostate cancer and bone stroma in an experimental model of androgen-independent prostate cancer bone metastasis. Cancer Res 61: 6012-6019

Milas L, Mason KA, Liao Z,Ang KK (2003) Chemoradiotherapy: emerging treatment improvement strategies. Head Neck 25: 152-167

Myers JN, Holsinger FC, Jasser SA, Bekele BN,Fidler IJ (2002) An orthotopic nude mouse model of oral tongue squamous cell carcinoma. Clin Cancer Res 8: 293-298

Nakayama J, Tahara H, Tahara E, Saito M, Ito K, Nakamura H, Nakanishi T, Tahara E, Ide T,Ishikawa F (1998) Telomerase activation by hTRT in human normal fibroblasts and hepatocellular carcinomas. Nat Genet 18: 65-68

Parkin DM, Bray F, Ferlay J,Pisani P (2005) Global cancer statistics, 2002. CA Cancer J Clin 55: 74-108

Peng XY, Won JH, Rutherford T, Fujii T, Zelterman D, Pizzorno G, Sapi E, Leavitt J, Kacinski B, Crystal R, Schwartz P,Deisseroth A (2001) The use of the L-plastin promoter for adenoviral-mediated, tumor-specific gene expression in ovarian and bladder cancer cell lines. Cancer Res 61: 4405-4413

Rodriguez R, Schuur ER, Lim HY, Henderson GA, Simons JW,Henderson DR (1997) Prostate attenuated replication competent adenovirus (ARCA) CN706: a selective cytotoxic for prostate-specific antigen-positive prostate cancer cells. Cancer Res 57: 2559-2563

Rothmann T, Hengstermann A, Whitaker NJ, Scheffner M,zur Hausen H (1998) Replication of ONYX-015, a potential anticancer adenovirus, is independent of p53 status in tumor cells. J Virol 72: 9470-9478

Shay JW, Werbin H,Wright WE (1996) Telomeres and telomerase in human leukemias. Leukemia 10: 1255-1261

Shay JW,Wright WE (2002) Telomerase: a target for cancer therapeutics. Cancer Cell 2: 257-265

Tan MH, Holyoke ED,Goldrosen MH (1977) Murine colon adenocarcinoma: syngeneic orthotopic transplantation and subsequent hepatic metastases. J Natl Cancer Inst 59: 1537-1544

Tsukuda K, Wiewrodt R, Molnar-Kimber K, Jovanovic VP,Amin KM (2002) An E2F-responsive replication-selective adenovirus targeted to the defective cell cycle in cancer cells: potent antitumoral efficacy but no toxicity to normal cell. Cancer Res 62: 3438-3447

Vokes EE, Crawford J, Bogart J, Socinski MA, Clamon G,Green MR (2005) Concurrent chemoradiotherapy for unresectable stage III non-small cell lung cancer. Clin Cancer Res 11: 5045s-5050s

Vokes EE, Weichselbaum RR, Lippman SM,Hong WK (1993) Head and neck cancer. N Engl J Med 328: 184-194

Wirth T, Zender L, Schulte B, Mundt B, Plentz R, Rudolph KL, Manns M, Kubicka S,Kuhnel F (2003) A telomerase-dependent conditionally replicating adenovirus for selective treatment of cancer. Cancer Res 63: 3181-3188

Hypoxia Induces Resistance to 5-Fluorouracil in Oral Cancer Cells via G1 Phase Cell Cycle Arrest

Sayaka Yoshiba*, Daisuke Ito, Tatsuhito Nagumo, Tatsuo Shirota, Masashi Hatori and Satoru Shintani

Department of Oral&Maxillofacial Surgery, School of Dentistry, Showa University, Tokyo145-8515, Japan.

*Address correspondence and reprint request to: Sayaka Yoshiba, Department of Oral&Maxillofacial Surgery, School of Dentistry, Showa University, Kitasenzoku 2-1-1, Ohta-ku, Tokyo145-8515, JAPAN.

Summary. Malignant tumors are exposed to various levels of hypoxic condition *in vivo*. To clarify the mechanism of the hypoxia-induced chemoresistance, we evaluated the effects of hypoxia on the resistance of oral squamous cell carcinoma (OSCC) cell lines to 5-fluorouracil (5-FU). OSCC cells were divided to two groups by the proliferation activity under hypoxic condition; hypoxia-resistant (HR) and hypoxia-sensitive (HS) cells. Growth of HS cells were inhibited by hypoxia and introduced to G1 arrest in cell cycle. 5-FU effect on HS cell viability was markedly reduced in hypoxic condition without an induction of chemoresistant related protein, P-glycoprotein. However, proliferation, cell cycle, and 5-FU sensitivity of HR cells were not affected by hypoxia. Hypoxia-inducible factor (HIF)-1α was induced by hypoxia in all OSCC cell lines, but diminished in HS cells within 48 hours. Expression of p21 and p27 was strongly augmented and CyclinD expression was reduced by hypoxia in HS cells. However, expression of these proteins was constitutive in HR cells during 48h hypoxic culture. Phosphorylation of mammalian target of rapamycin (mTOR) was reduced by hypoxia in HS cells. From these findings, we concluded that HS OSCC cells acquire 5-FU resistance under hypoxia by G1/S transition through an upregulation of cell cycle inhibitors.

Key words. 5-FU, oral cancer, resistance, hypoxia, cell cycle, G1 phase

Introduction

Malignant tumors are exposed to various levels of hypoxic condition *in vivo* (Brown and Wilson 2004). Hypoxia has been known to be associated with malignant phenotypes of tumor cells and prognosis of the patients (Janssen et al. 2005). Recently it has been reported that tumor cells under a certain level of hypoxia are resistant to chemotherapies (Brown 2000, Sakata et al. 1991, Song et al. 2006, Teicher 1994). To clarify the mechanism of the hypoxia-induced chemoresistance of oral cancer, we evaluated the effects of hypoxia on the resistance of oral squamous cell carcinoma (OSCC) cell lines to 5-fluorouracil (5-FU) analyzed the molecular mechanisms of acquired chemoresistance.

Materials and Methods

Materials

Oral squamous cell carcinoma cell lines HSC3, CA9-22, NA, and SAS were used. Hypoxic condition was controlled by an Invivo2 humidified airtight O_2-control incubator (Ruskinn Technology, Leeds, United Kingdom).

Methods

Cell proliferation, viability and 5-FU sensitivity were evaluated by crystal violet mitogenic assay.
The percentages of cells in the sub-G_1, G_0/G_1, S, and G_2/M phases were evaluated by DNA content analysis. The protein expression were analyzed by Western blot analysis.

Results

1. OSCC cells were divided into two groups according to the proliferation rate under hypoxic condition; hypoxia-resistant (HR) and hypoxia-sensitive (HS) cells (Fig. 1A & B).
2. HS cell lines were introduced to G1-phase cell cycle arrest under hypoxia (Fig. 2).
3. Hypoxic culture resulted in a decrease of 5-FU sensitivity in HS cells; but not in HR cells (Fig. 3).
4. Hypoxia did not induce the P-gp expression in both HS and HR cells (Fig. 4).
5. The expression of p21 and p27 was augmented after 48hr hypoxic culture in HS cells (Fig. 5).
6. mTOR was hypophosphorylated in HS cells under the hypoxic condition (Fig. 6A).
7. The expression levels of AMPK, p-AMPK and REDD1 were not affected by hypoxia.(Fig. 6B)

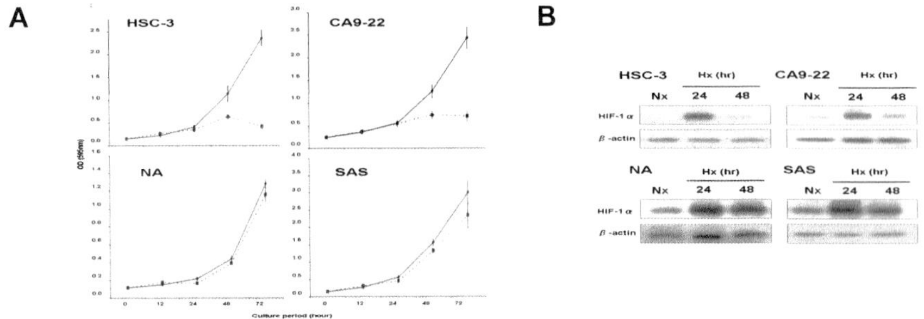

Fig. 1 Growth of OSCC cells and the expression of HIF-1α in OSCC cells under hypoxic condition. (A) Cells were plated on 96-well plates (3 x 10^3 cells per well) and cultured for 12-72 hours under normoxic (solid lines) or hypoxic (dotted lines) condition. Proliferation of HSC-3 and CA9-22, but NA and SAS, is suppressed under hypoxic culture. Cell viability was examined by crystal violet mitogenic assay. The data indicate the mean ± SD of triplicate wells. Representative data from three individual experiments are shown. *; $p<0.01$ vs normoxia, **; $p<0.001$ vs normoxia. (B) Cells were cultured in normoxia (Nx) or hypoxia (Hx) for 24-48 hours and the expression of HIF-1α was examined by Western blot analysis. HIF-1α expression is induced by hypoxia, but diminished in HS cells within 48 hours of culture.

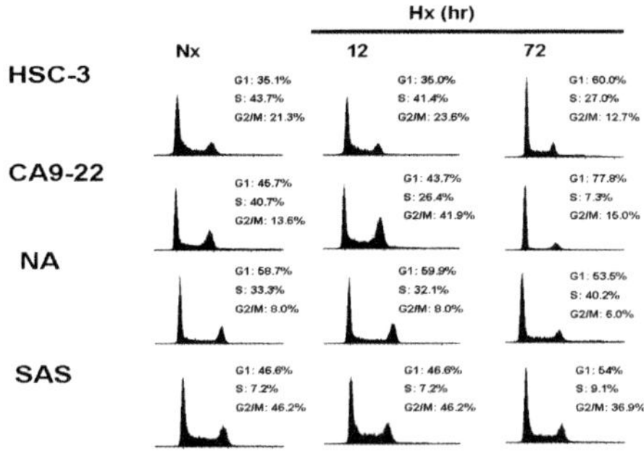

Fig. 2. Cell cycle of OSCC cells under hypoxia. (Hypoxia induces G1 phase cell cycle arrest in HS cells.) Cells were cultured in normoxia (Nx) or hypoxia (Hx) for 12-72 hours and DNA content analyses were performed by PI staining and subsequent flow cytometry. From the results of DNA histograms, the percentage of cells in G1, S, and G2/M phases were evaluated.

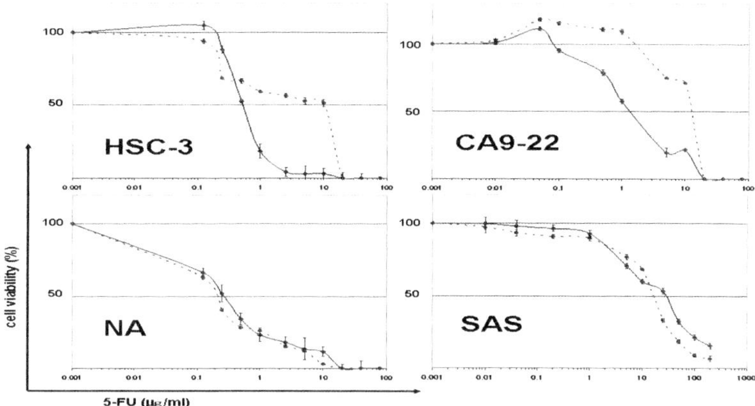

Fig. 3 Chemosensitivity of OSCC cells to 5-FU under hypoxia. (Hypoxia induces insensitivity to 5-FU in HS cells.) Cells were cultured in normoxia (Nx) or hypoxia (Hx) for 72 hours in the presence of 0-1000μg/ml of 5-FU. Cell viability was examined by crystal violet mitogenic assay as described in *Materials and Methods*. The data indicate the mean ± SD of triplicate wells.

As shown in Fig.3, less than 0.1 μg/ml of 5-FU displayed practically no effect on the growth of hypoxia-sensitive HSC-3 and CA9-22 cells under normoxia, and most of the cells were not survived in the 5-FU concentration of more than 10 μg/ml. Calculated 50%-inhibitory concentration (IC50) values of 5-FU were approximately 0.6 μg/ml for HSC-3 and 1.5 μg/ml for CA9-22. Hypoxic culture of these cells resulted in a decrease in 5-FU sensitivity; IC50 values turned in hypoxia to 10 and 15 μg/ml for HSC-3 and CA9-22, respectively. On the other hand, 5-FU sensitivity of HR cells was not changed in hypoxia. IC50 values of 5-FU were 0.3 μg/ml for NA and 13 μg/ml for SAS in normoxia, 0.2 μg/ml for NA and 12 μg/ml for SAS in hypoxia.

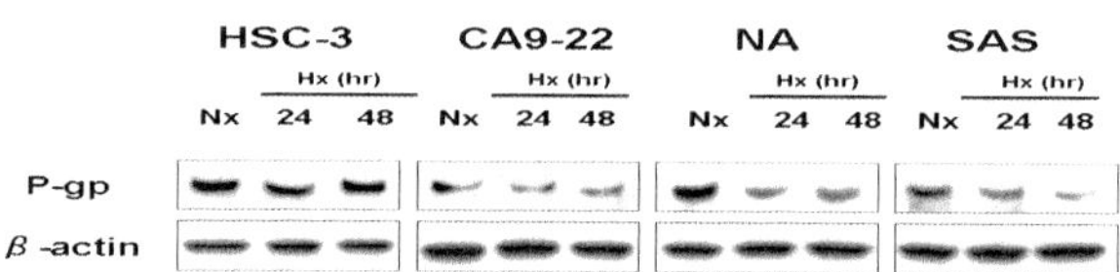

Fig. 4 Expression of p-glycoprotein in OSCC cells under hypoxia. (Expression of p-glycoprotein is not altered by hypoxia.) Cells were cultured in normoxia (Nx) or hypoxia (Hx) for 24-48 hours and the expression of p-glycoprotein (p-gp) was examined by Western blot analysis.

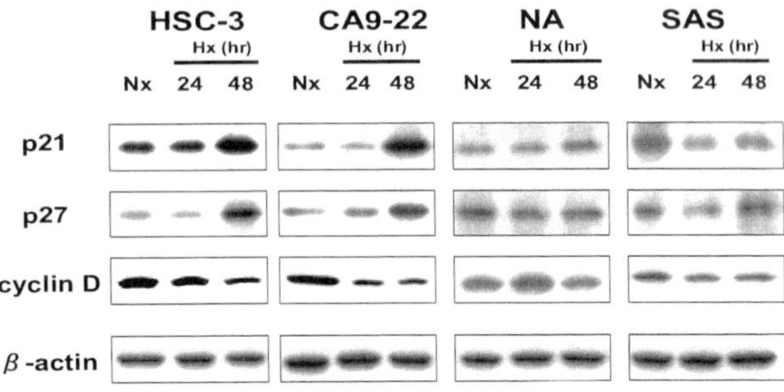

Fig. 5 Expression of cell cycle regulators in OSCC cells under hypoxia. (Hypoxia induces marked expression of p21 and p27 in HS cells.) Cells were cultured in normoxia (Nx) or hypoxia (Hx) for 24-48 hours and the expression of p21, p27, and cyclin D was examined by Western blot analysis. Expression of p21 and p27 in HS cells under normoxia was comparable to that in HR cells. After 48-hour of hypoxia, their expression was upregulated in HS cells but not in HR cells. Constitutive expression of cyclin D was observed in all the cell lines examined and did not change the expression levels in HR cells during 48 hours hypoxic culture.

Discussion

We investigated the mechanism of hypoxia-induced 5-FU resistance of OSCC. The reason why cells more sensitive to hypoxic stress are more capable to acquire drug resistance remains unclear. We hypothesize that HS cells block their own proliferation activity in response to various stresses to save their energy and intercept the deleterious outside effects like anticancer drugs. It is possible according to our present findings that unknown cascades or pathways are associated. There is a missing link between HIF-1 α induction by energy starvation stress and G1 arrest via mTOR hypophosphorylation. Candidate molecules for novel anticancer gene targeting are expected be discovered through elucidating the detailed molecular mechanisms of cellular stress responses.

In addition, we have to make clear the resistance mechanism of other chemotherapeutic agents in further studies.

A

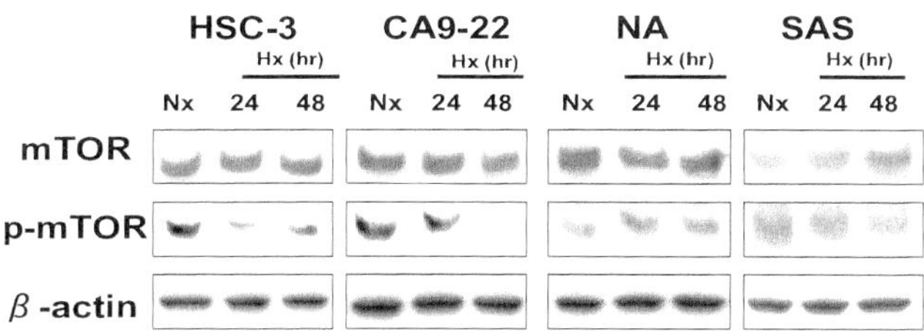

B

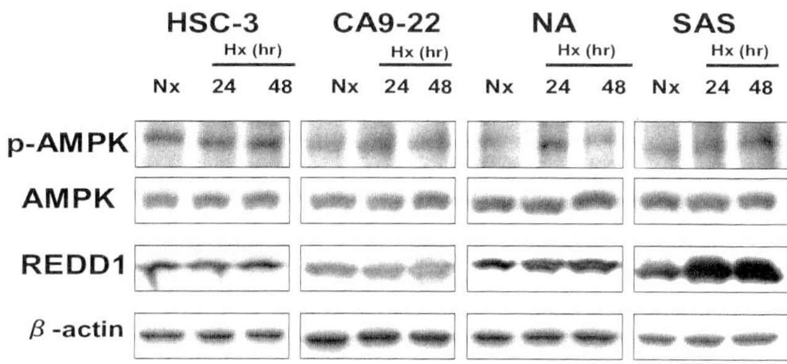

Fig. 6 mTOR expression and phosphorylation in OSCC cells under hypoxia. And the expression AMPK, REDD and phosphorylation of AMPK in OSCC cells under hypoxia. (A) Cells were cultured in normoxia (Nx) or hypoxia (Hx) for 24-48 hours and the expression of mTOR and tyrosine-phosphorylated mTOR was examined by Western blot analysis. mTOR is dephosphorylated in HS cells under hypoxia. (B) Cells were cultured in normoxia (Nx) or hypoxia (Hx) for 24-48 hours and the expression of AMPK, REDD1, and tyrosine-phosphorylated AMPK was examined by Western blot analysis. Hypoxia does not affect the expression of AMPK/REDD1 and AMPK phosphorylation status. The results shown in Figs. 6 suggested that induction of p21 and p27 CKIs in HS cells might be mediated by hypophosphorylation and subsequent inactivation of mTOR, and that AMPK and REDD1 might not be associated with this pathway.

References

Brown JM,Wilson WR (2004) Exploiting tumour hypoxia in cancer treatment. Nat Rev Cancer 4: 437-447

Brown JM (2000) Exploiting the hypoxic cancer cell: mechanisms and therapeutic strategies. Mol Med Today 6: 157-162

Janssen HL, Haustermans KM, Balm AJ,Begg AC (2005) Hypoxia in head and neck cancer: how much, how important? Head Neck 27: 622-638

Sakata K, Kwok TT, Murphy BJ, Laderoute KR, Gordon GR,Sutherland RM (1991) Hypoxia-induced drug resistance: comparison to P-glycoprotein-associated drug resistance. Br J Cancer 64: 809-814

Song X, Liu X, Chi W, Liu Y, Wei L, Wang X,Yu J (2006) Hypoxia-induced resistance to cisplatin and doxorubicin in non-small cell lung cancer is inhibited by silencing of HIF-1alpha gene. Cancer Chemother Pharmacol 58: 776-784

Teicher BA (1994) Hypoxia and drug resistance. Cancer Metastasis Rev 13: 139-168

Key Word Index